Conference Proceedings of the Society for Experimental Mechanics Series



For other titles published in this series, go to
www.springer.com/series/8922

Tom Proulx
Editor

# Nonlinear Modeling and Applications, Volume 2

Proceedings of the 28th IMAC, A Conference on Structural Dynamics, 2010

Springer

*Editor*
Tom Proulx
Society for Experimental Mechanics, Inc.
7 School Street
Bethel, CT 06801-1405
USA
tom@sem1.com

ISSN 2191-5644 e-ISSN 2191-5652
ISBN 978-1-4614-2890-9 ISBN 978-1-4419-9719-7 (eBook)
DOI 10.1007/978-1-4419-9719-7
Springer New York Dordrecht Heidelberg London



Printed on acid-free paper

Springer is part of Springer Science+Business Media (www.springer.com)

# Preface

*Nonlinear Modeling and Applications* represents one of five clusters of technical papers presented at the 28th IMAC, A Conference and Exposition on Structural Dynamics, 2010 organized by the Society for Experimental Mechanics, and held at Jacksonville, Florida, February 1-4, 2010. The full proceedings also include volumes on Structural Dynamics and Renewable Energy, Dynamics of Bridges, Dynamics of Civil Structures and, Structural Dynamics.

Each collection presents early findings from experimental and computational investigations on an important area within Structural Dynamics. The current volume on *Nonlinear Modeling and Applications* includes studies on Nonlinear Modeling, Nonlinear System Identification, and Nonlinear Experimentation

Nonlinearities in structural dynamic systems are increasingly important in advanced applications involving biological systems, advanced aircraft, complex ground vehicles, micro/nano-systems, and other structures. Nonlinear Modeling and Applications addresses a range of topics in nonlinear structural dynamics including analytical modeling and simulation, experimentation and system identification, and design, health monitoring, and control.

The organizers would like to thank the authors, presenters, session organizers and session chairs for their participation in this track. The Society would like to thank the organizers of this Focus Topic, D.E. Adams, *Purdue University*; G. Kerschen, *University of Liege* for their efforts.

Bethel, Connecticut

Dr. Thomas Proulx
Society for Experimental Mechanics, Inc

# Contents

Proceedings of the IMAC-XXVIII
February 1–4, 2010, Jacksonville, Florida USA


# Dynamics of a System of Coupled Oscillators with Geometrically Nonlinear Damping

**D.K. Andersen** [(1), (2)], **A.F. Vakakis**[(3)], **L.A. Bergman**[(4)]

(1) Graduate Student, Department of Mechanical Science and Engineering,
University of Illinois at Urbana-Champaign, Urbana, IL, U.S.A. Email: dandersn@illinois.edu
(2) Engineer, Defense Nuclear Facilities Safety Board, 625 Indiana Ave. NW, Suite #700,
Washington D.C., 20004, U.S.A. Email: davida@dnfsb.gov
(3) Professor, Department of Mechanical Science and Engineering, University of Illinois at Urbana-Champaign,
3003 Mechanical Engineering Laboratory, 1206 W. Green Street, MC-244, Urbana, IL, 61801, U.S.A.
Email: avakakis@illinois.edu
(4) Professor, Department of Aerospace Engineering, University of Illinois at Urbana-Champaign,
321e Talbot Laboratory, 104 S. Wright Street, Urbana, IL, U.S.A, 61801. Email: lbergman@illinois.edu

**ABSTRACT**

The influence of adding a geometrically nonlinear viscous damper to a system of coupled oscillators with essential nonlinear stiffness will be discussed. All nonlinear terms are restricted to the coupling terms between a linear oscillator and light attachment. We show that the addition of the nonlinear damper introduces dynamics not observed with linear damping. In fact, we find the surprising result that the nonlinear damper introduces new dynamics into the problem, and its effect on the dynamics is far from being purely parasitic - as one would expect in the case of weak linear viscous dissipation. Similar to essential nonlinear stiffness, geometrically nonlinear damping of the type considered in our work is physically realizable by means of linear viscous damping elements. Numerical work examining this problem will be discussed.

## 1 INTRODUCTION

A two-mass oscillating system was shown to exhibit complicated dynamics when a nonlinear stiffness element coupled the motion of a linear oscillator (LO) and a light attachment [1,2,3]. These works showed that through the excitation of the impulsive orbits of the system, a large portion of the system energy could be pumped from the impulsively excited LO to the attachment and then dissipated. Due to the efficacy of the light attachment to attract and damp out a significant portion of the system energy, it is referred to as the nonlinear energy sink (NES). This work has been extended by changing the coupling damper from linear to nonlinear. The quadratic coupling damper and cubic coupling stiffness are physically realizable by aligning a linear spring and a linear damper parallel to each other and perpendicular to the direction of motion.

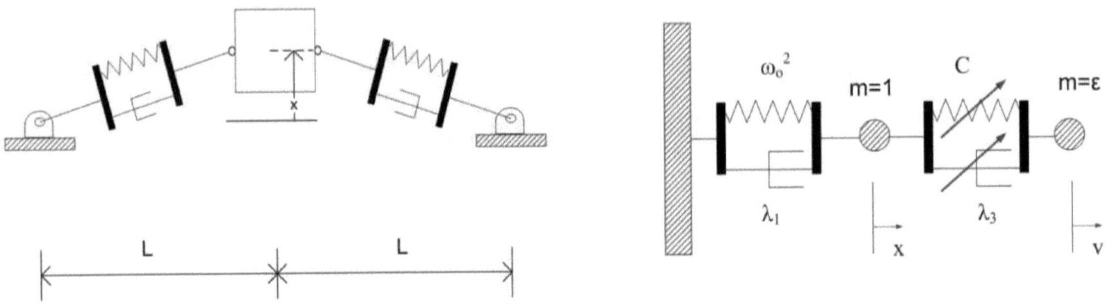


(a) (b)

**Figure 1: (a) Physical Realization of Nonlinear Damping and Nonlinear Stiffness (b) Idealization of Two-Mass System**

## 2 SYSTEM DEFINITION

Different approaches for physically realizing quadratic damping and cubic stiffness were provided in [4,5]. The configuration discussed in [5] can be seen in Figure 1(a). The general system of equations for a two-mass system with nonlinear coupling terms is expressed as follows:

$$\begin{aligned} &\ddot{x}+\lambda_1\dot{x}+\lambda_2\left(\dot{x}-\dot{v}\right)+\lambda_3\left(x-v\right)^2\left(\dot{x}-\dot{v}\right)+\omega_o^{\,2}x+C\left(x-v\right)^3=0 \\ &\varepsilon\ddot{v}+\lambda_2\left(\dot{v}-\dot{x}\right)+\lambda_3\left(v-x\right)^2\left(\dot{v}-\dot{x}\right)+C\left(v-x\right)^3=0 \end{aligned} \tag{1}$$

The analyzed system was impulsively loaded by imposing the following initial conditions:

$$x(0)=v(0)=\dot{v}(0)=0,\ \dot{x}(0)=I_o \tag{2}$$

For all numerical work presented in this paper, the system coefficients used are:

$$\varepsilon=0.05,\ \lambda_1=0.0015,\ \lambda_2=0,\ \lambda_3=0.010,\ C=1.0 \tag{3}$$

The idealization of the system defined by eqs. 1-3 is shown in Figure 1(b).

The system was solved for a 0.1 second time step with the ODE45 numerical differential equation solver of MATLAB [6]. Prior to examining the detailed dynamics of the system response, a parametric study was performed to determine the LO system that would result in the optimal NES energy dissipation. The energy dissipated in the NES attachment from t = 0 to $t = t_f$ can be expressed as follows:

$$E_{dis}(\%)=\frac{\int_0^{t_f}\left\{\lambda_2+\lambda_3\left[v(t)-x(t)\right]^2\right\}\left[\dot{v}(t)-\dot{x}(t)\right]^2 dt}{I_o^{\,2}/2}\ x\ 100\% \tag{4}$$

The results from the parametric study are provided in Figure 2, with the system parameters of LO impulse ($I_o$) and LO natural frequency ($\omega_o$) varied. From this figure, it is seen that the maximum amount of the input energy (>95%) is dissipated in the NES when the LO natural frequency exceeds 2.0 rad/sec. For various values of $\lambda_3$, including $\lambda_3$=0.01 in Figure 2, a LO natural frequency of 2.9 rad/sec consistently produced optimal NES energy dissipation. Thus, this LO frequency will be used for all subsequent numerical work described in this paper. The system with $\omega_o$ = 1.0 rad/sec was also examined in detail, but has less interesting dynamics than the system with $\omega_o$ = 2.9 rad/sec.

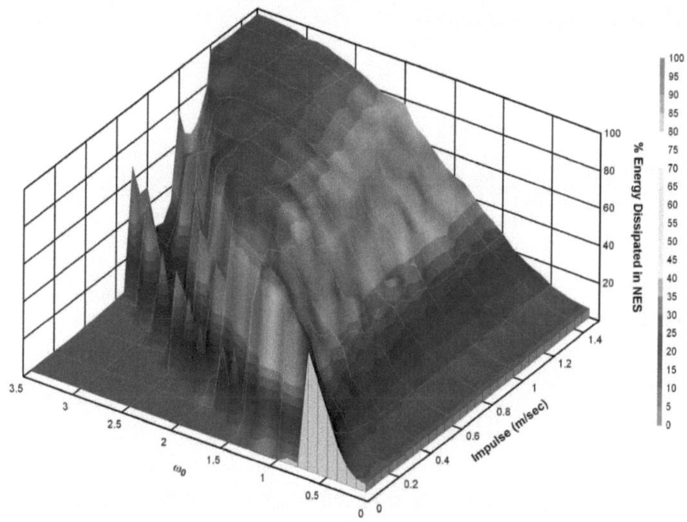


**Figure 2: Energy dissipation in the NES for varying values of LO natural frequency ($\omega_o$) and impulsive loading magnitude ($I_o$), $\varepsilon = 0.05$, $\lambda_1 = 0.0015$, $\lambda_2 = 0$, $\lambda_3 = 0.010$, $C = 1.0$**

## 3 NONLINEAR NORMAL MODES AND FREQUENCY ENERGY PLOTS

The time history solutions from the numerical simulation were post-processed using the wavelet transform. Wavelet transforms were performed using the MATLAB tool developed at the University of Liege [7]. Wavelet transforms for relative displacement are plotted on Frequency Energy Plots (FEPs) and compared to the nonlinear normal mode (NNM) backbone curves of the Hamiltonian system. The FEPs were defined in [1,3] and developed for the cubic coupling stiffness system. Wavelet transforms are a moving Fourier transform of frequency versus time; for the FEP the wavelet frequency is instead plotted against instantaneous system energy. Hence, the progression on the FEP from high energy to low energy corresponds to a forward progression in time as the damper elements dissipate energy from the system. It was necessary to recompute the system backbone curves for the study in this paper since the LO natural frequency is changed from $\omega_o = 1.0$ rad/sec to $\omega_o = 2.9$ rad/sec. The Hamiltonian system NNMs are computed using the non-smooth transformations developed by Pilipchuk in [8]. This transformation is:

$$\begin{aligned} v(t) &= e(t/\alpha)y_1\left[\tau(t/\alpha)\right] \\ x(t) &= e(t/\alpha)y_2\left[\tau(t/\alpha)\right] \end{aligned} \tag{5}$$

Where:

$$\begin{aligned} \tau(u) &= \frac{2}{\pi}\sin^{-1}\left(\sin\frac{\pi}{2}u\right) \\ e(u) &= \dot{\tau}(u) \end{aligned} \tag{6}$$

$\tau(u)$ corresponds to the saw-tooth function and $e(u)$ corresponds to the square wave. Their graphical depiction is presented in [1,3,8].

The non-smooth transformation is applicable to systems ranging from a linear oscillator to a strongly nonlinear vibro-impact oscillator [8]. After applying 'smoothing conditions' that eliminate singular terms, the transformed system of equations for the Hamiltonian is written as [1,3]:

$$
\begin{aligned}
y_1' &= y_3 \\
y_2' &= y_4 \\
y_3' &= -\frac{C}{\varepsilon}\alpha^2 (y_1 - y_2)^3 \\
y_4' &= -\omega_o^2\alpha^2 y_2 - C\alpha^2 (y_2 - y_1)^3
\end{aligned} \tag{7}
$$

To formulate the NNM, displacement boundary conditions are imposed:

$$
\begin{aligned}
y_1(-1) = y_1(+1) = 0 \\
y_2(-1) = y_2(+1) = 0
\end{aligned} \tag{8}
$$

In eq. 7 $\alpha$ represents an unknown quarter-period for a NNM solution, or equivalently it can be viewed as the NNM eigenvalue. For a given value of $\alpha$, the unknown velocity boundary conditions can be solved for using a shooting method. The nature of the velocities at $\tau = -1$ and $\tau = +1$ determine whether a modal solution should be characterized as symmetric (S) or unsymmetric (U). A restatement of the naming conventions used in [1,2,3] is provided as follows:

i. *Symmetric periodic orbits*. Snm±

Symmetric orbits are NNMs where:

$$
\begin{aligned}
y_1'(-1) = \pm y_1'(+1) \\
y_2'(-1) = \pm y_2'(+1)
\end{aligned} \tag{9}
$$

For $-1 \le \tau \le +1$, the index *n* corresponds to the number of half-waves in $y_1$ and the index *m* corresponds to the number of half-waves in $y_2$. Symmetric orbits Snm+ corresponds to in-phase modes and Snm- corresponds to out-of-phase modes. For example S11- corresponds to the modal response where $y_1$ (and v) and $y_2$ (and x) have a 1:1 out-of-phase, internal resonance. As another example, S15+ corresponds to the modal response of a 1:5 in-phase, internal resonance where the LO oscillates at 5 times faster frequency than the NES. Lastly, it is important to note that S(k*n)(k*m) modes, where k is a positive integer, plot as equivalent curves to Snm on the FEP.

ii. *Unsymmetric periodic orbits*. Upq

Unsymmetric orbits are NNMs where the conditions of eq . 9 are not satisfied. The orbits U(m+1)m are those which bifurcate from the S11- branch at $\omega \approx (m+1)\omega_o/m$. For example, U43 will bifurcate from S44- (equivalently S11- on the FEP) at $\omega \approx 4/3\ \omega_o$. There exist modal solutions on the unsymmetric tongues that satisfy the additional conditions:

$$
\begin{aligned}
y_1'(-1) = 0 \\
y_2'(-1) \neq 0
\end{aligned} \tag{10}
$$

These modes are called the *special periodic orbits* (SPO) and represent the impulsive excitation of the LO mass. The SPO manifold is identified on the FEP as the green dotted curve (not to be confused with the black, vertical dashed line which indicates the input energy into the system).

## 4 NUMERICAL WORK

Numerical simulation was performed on the system defined by eqs. 1-3. A parametric study was performed by varying the initial LO impulse value, $I_o$, from 0.05 to 1.5 at increments of 0.05. For values of $I_o$ ranging from 0.05 to 0.45 the system response is solely S11-. When the impulse is increased to $I_o$ = 0.50 a bifurcation occurs from S11- to S13. A similar bifurcation is observed for $I_o$ values of 0.55, 0.60 and 0.65. The time history and FEP for $I_o$ = 0.65 are presented in Figure 3(a) and 3(b), respectively, and show sustained dynamics in the NES and the S11- to S13 bifurcation.

The time history response and FEP for $I_o$ = 0.70 are shown in Figure 4(a) and (b). The first behavior to note on the FEP is that the system bifurcates from S11- resonance capture to S13, and then jumps back up in frequency to S12. This bifurcation occurs once the input energy exceeds the energy of the S12 SPO. Also, a beating phenomenon is observed on the FEP after bifurcation. In previous work [1,2,3] beats were observed to initiate sub-harmonic resonance capture, but not follow it. Lastly, the time history plot again shows the dynamics of the NES sustaining even after 10,000 seconds of simulation. Evaluation of the linearly damped system with identical system parameters, except for the values of $\lambda_2$ and $\lambda_3$ reversed, showed the system dynamics to dissipate by 1,000 seconds.

The FEP for $I_o$ values of 0.75 and 0.85 are shown in Figures 5(a) and (b). In [1,2,3] it was shown that the dynamics of the linearly damped system ($\lambda_2 \neq 0$, $\lambda_3 = 0$) followed the backbone curves of the FEP. However, contrary to this behavior, the nonlinear damped system ($\lambda_2 = 0$, $\lambda_3 \neq 0$) for the larger impulse values of Figure 5 results in dynamics which deviate from the FEP backbone curves. This deviation can be seen in the transition from S11+ resonance capture to the S13 bifurcation. It is also important to note that the system bifurcates from S11+ down to S13, then up to S12, then back down to S13. The S12 and S13 bifurcations are not possible until the input energy (vertical dashed line) exceed the energy level of the SPO for S12 and S13 NNMs, respectively. These occurrences are represented by the three way intersection of the input energy (vertical, black dashed line), the SPO invariant manifold (dotted green curve), and the S12 or S13 backbone curves. Lastly, it is important to note that the response again exhibits a post-bifurcation beating.

Based on the observations above, the quadratically damped system shows interesting dynamics that warrant further analytical and experimental evaluation.

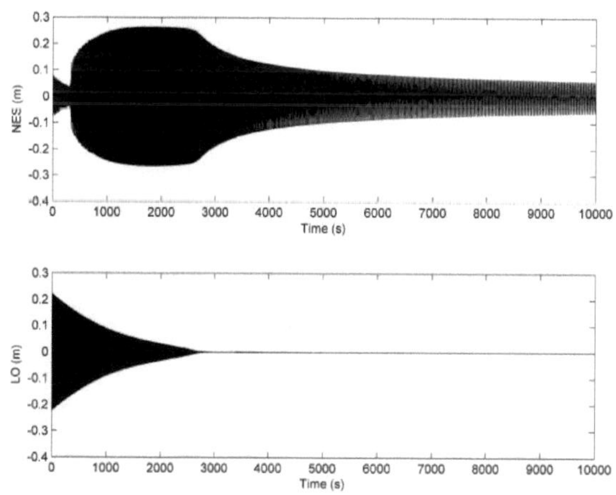


(a)

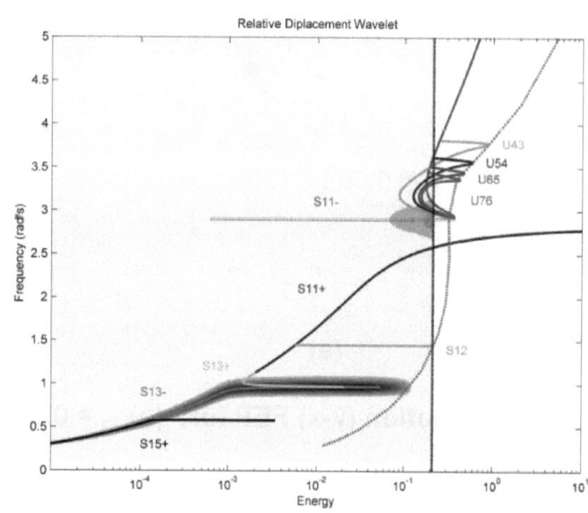


(b)

**Figure 3: (a) Time history response for NES and LO, $I_o$ = 0.65**
**(b) Relative motion (v-x) FEP for $I_o$ = 0.65**

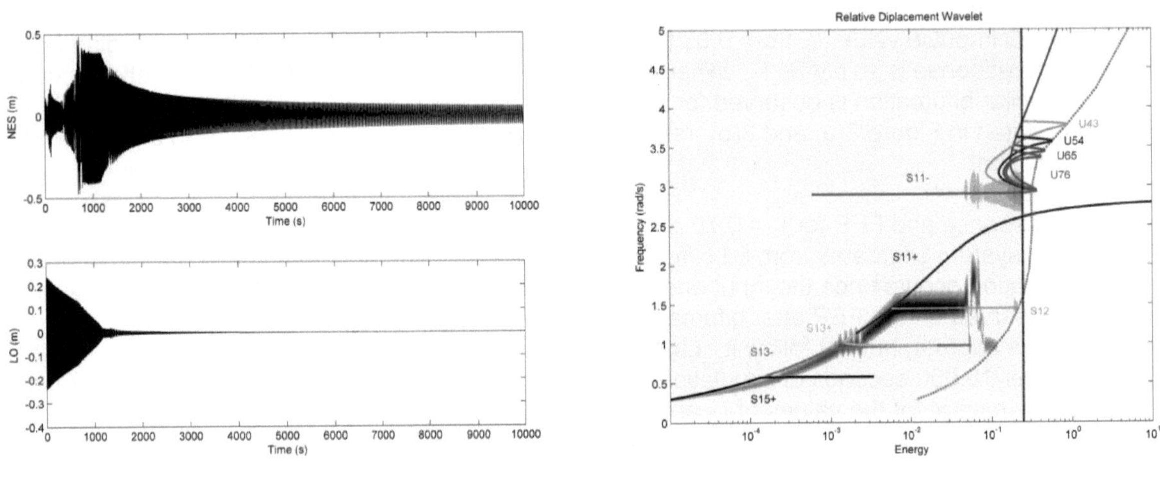


**(a)** **(b)**

**Figure 4: Response for $I_o$ = 0.70, (a) Time histories (b) (v-x) FEP**

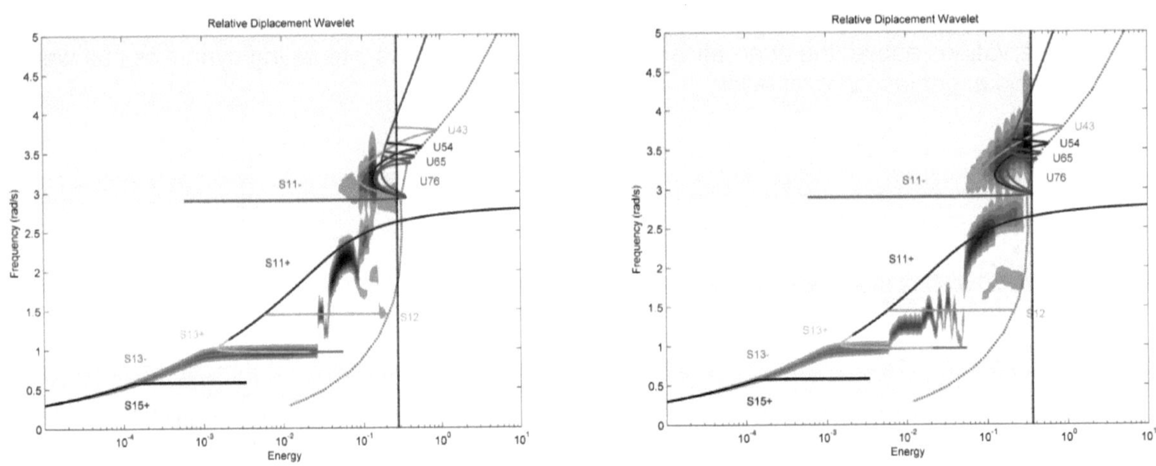


**(a)** **(b)**

**Figure 5: Relative motion (v-x) FEP for: (a) $I_o$ = 0.75 (b) $I_o$ = 0.85.**

## 5 CONCLUDING REMARKS

Numerical work has provided evidence that interesting dynamics exists for a system with nonlinear coupling stiffness and a nonlinear coupling damper. Observations to note for the quadratically damped/cubic stiffness system are:

- The system response deviates significantly from the NNM backbone curves for certain impulse values. The quadratic damper appears to affect the system dynamics and not be only parasitic in nature. This would indicate a need to revise Hamiltonian backbones curves to account for the new dynamics.
- The system responds with post-bifurcation beating for certain loading cases.
- A strong bifurcation from S11- down to S13, back up to S12, and then back down to S13 occurs for certain loading cases.
- The NES responds for large time durations when compared to the linearly damped system.

The work documented in this paper provides a solid basis for continuing the examination of this system. Analytical and experimental work is planned for the near future.

## 6 DISCLAIMER

The views expressed in this paper are solely those of the authors and no endorsement by the Defense Nuclear Facilities Safety Board is intended or should be inferred.

## REFERENCES

[1] Lee et al., Complicated Dynamics of a Linear Oscillator with a Light, Essentially Nonlinear Attachment, PhysicaD, 204 (2005), p. 41-69.

[2] Kerschen et al., Irreversible Passive Energy Transfer in Coupled Oscillators with Essential Nonlinearity, SIAM Journal of Applied Mathematics, Vol. 66, No. 2, p. 648-679.

[3] Vakakis et al., Nonlinear Targeted Energy Transfer in Mechanical and Structural Systems I, Springer, 2008.

[4] Koumousis, V. and Kefala, G., On the Dynamic Behavior of a Lightweight Isolator for Museum Artifacts, International Conference on Structural Engineering Dynamics, 2009.

[5] Triplett et al., Energy Harvesting from an Impulsive Load with Essential Nonlinearities, ASME International Design Engineering Technical Conference, Paper DETC2009-86669, 2009.

[6] MATLAB Software, Version 7.9, R2009b.

[7] Lenaerts, V. and Argoul, P., MATLAB Wavelet Transform package, 2001.

[8] Pilipchuk, V.N., Analytical Study of Vibrating Systems with Strong Non-Linearities by Employing Saw-Tooth Time Transformations, Journal of Sound and Vibration, 192(1), p. 43-64.

Proceedings of the IMAC-XXVIII
February 1–4, 2010, Jacksonville, Florida USA


# Assigning the Nonlinear Distortions of a Two-input Single-output System

W. D. Widanage, J. Schoukens
Vrije Universiteit Brussel, Department ELEC, Pleinlaan 2, B1050 Brussels, Belgium

## Abstract

In a multiple-input multiple-output (MIMO) nonlinear system the nonlinear contributions from each input to each output may vary significantly. The nonlinear distortions are composed of contributions that are either purely due to each individual input or a combination with some of the inputs. A three stage experimental design method valid for a wide range of nonlinear systems is presented that detects and classifies, in the frequency domain, the level of these nonlinear contributions. For systems that require nonzero operating points, the contribution from each input and their combinations is conditioned by the operating levels. Periodic broadband excitation signals with several harmonics suppressed are used as the inputs to reduce the noise contributions and evaluate the nonlinear distortion levels present at the suppressed harmonics. Alternatively, each input signal can be designed with a harmonic specification such that the harmonics of the output signal indicate the presence of these nonlinear contributions. As a single experiment technique it requires less time for measurements, however the input harmonics become very sparse as the order of the nonlinearity increase that the signals becomes impractical for experimental use. Experimental results for a two-input single-output system are presented demonstrating the effectiveness of the techniques.

## Nomenclature

| | |
|---|---|
| $u(t)$ and $v(t)$: | Input signals |
| $U(k)$ and $V(k)$: | Discrete Fourier transforms of $u(t)$ and $v(t)$ |
| $M$: | Number of samples per signal period |
| $G_{\alpha U_{\alpha-j}V_j}$: | $\alpha^{th}$ order frequency domain Volterra kernel |
| $\alpha$ : | Highest nonlinear order |
| $F$: | Total number of harmonics |
| $D_1(k), D_1^a(k)$: | Nonlinear distortions at harmonic $k$ from input $u(t)$ |
| $D_2(k), D_2^b(k)$: | Nonlinear distortions at harmonic $k$ from input $v(t)$ |
| $D_3(k), D_3^c(k)$: | Nonlinear distortions at harmonic $k$ from both inputs |
| $SS_u, SS_v, SS_{uv}$: | Sets of harmonics indicating nonlinear distortions from either input $u(t)$, $v(t)$ or both |

## 1 Introduction

A linear approximation of a nonlinear system is its frequency response estimate. When using broadband excitations to estimate a frequency response, the harmonics of the output signal are affected by both noise contributions and inter-harmonic modulations of the input harmonics due to the nonlinear effect. For a single-input single-output (SISO) system, with no external noise entering the system at its input, the inter-harmonic modulations can only be a function of the input harmonics. For a two-input singe-output (TISO) system the inter-harmonic modulations appearing at the output signal are functions of harmonics from that of each input and their combinations.

The choice of the input signals influences the accuracy of the frequency response estimate. For SISO systems the search for input harmonics where there are no effects of inter-harmonic combinations by non-linearities, appearing at these harmonics have been studied and such signals are known as no inter-harmonic distortion (NID) multisines [1]. The harmonics of such signals become sparse for increasing nonlinear order becoming impractical for experimental use. In reference [2], orthogonal random phase multisines are shown to outperform Gaussian noise, periodic noise and random phase multisines when estimating frequency responses of a MIMO system, while in [3] they are shown to outperform three multisine signals when estimating linear MIMO systems with output noise.

In this paper two procedures are described whereby the first method detects and classifies the level of the nonlinearities and the second classifies the nonlinearities as contributions from each input and their combination. The fist Section describes the type of signals used and the assumptions imposed on the nonlinear MIMO system, followed by the description of the two methods; the experimental and simulation results of the two procedures and concluding remarks.

# 2 Theoretical Preliminaries

## 2.1 Input signals

In this work the input signals used to excite the system are broadband periodic signals. Such signals offer several advantages [4]. Being periodic, leakage effects can be eliminated by measuring an integer multiple of periods and the effect of measurement noise can be reduced by averaging over periods [5]. The type of periodic signals used in this work are the *random phased multisine* signals.

$$u(t) = \sum_{k=-M/2}^{M/2} U(k) e^{j2\pi k f_s t/M} \tag{1}$$

$$U(k) = A(k) e^{j\phi_k} \tag{2}$$

Such signals allow arbitrary harmonics to be suppressed by setting $U(k) = 0$ while assigning a desired amplitude (power) level in the remaining harmonics. In equation (2) $A(k)$ is the real valued amplitude and $\phi_k$ is the phase at harmonic $k$. The phase is an identically and uniformly distributed random variable on the interval $[0, 2\pi)$. Further $U(k)$ is a complex variable with $U(-k) = U^*(k)$, and $f_s$ is the sampling frequency and $M$ the number of samples per period.

The generation of additional frequencies is an example of a nonlinear phenomenon [6]. This effect allows the level of the nonlinearity to be detected. By suppressing certain harmonics in the input spectrum, the resulting amplitude levels in the output spectrum at the suppressed harmonics indicates the presence and level of the nonlinear distortions [7, 8]. Further if all even harmonics (including the d.c) are suppressed, the distortions at the even harmonics are due to even order nonlinearities and if some odd harmonics are suppressed, the distortions at the suppressed odd harmonics are due to odd order nonlinearities [9].

The signals reported in this work, have all even harmonics suppressed and several randomly selected odd harmonics suppressed.

## 2.2 Nonlinear system

It is assumed that the nonlinear system can be modelled as a Volterra series [6] in the mean square sense, with the error converging to zero as the kernel order tends to infinity. Volterra series covers a wide variety of nonlinear systems and is widely used to describe SISO nonlinear systems. With periodic input signals, the Volterra series in the frequency domain extended to a TISO is:

$$Y(k) = \sum_{\alpha=1}^{\infty} \sum_{j=0}^{\alpha} \sum_{\substack{l_1,\ldots,l_{\alpha-j-1}=-M/2 \\ p_1,\ldots,p_{j-1}=-M/2}}^{M/2} G_{\alpha U_{\alpha-j} V_j}(l_1,\ldots,l_{\alpha-j-1},p_1,\ldots,p_{j-1}) \prod_{i=1}^{\alpha-j} U(l_i) \prod_{i=1}^{j} V(p_i) \tag{3}$$

In equation (3) the outer summation is for the kernel order, the next sums over the combinations of $U(k)$ and $V(k)$ for a given order and the inner summation sums the frequencies. For a given order and combination of inputs, the $\alpha$th order frequency kernel is given by $G_{\alpha U_{\alpha-j} V_j}$ and frequency $k = \sum_{i=1}^{\alpha-j} l_i + \sum_{i=1}^{j} p_i$. A kernel of the form $G_{\alpha U_\alpha V_0}$ indicates that the input signal $v(t)$ has no influence on the output signal and similarly $G_{\alpha U_0 V_\alpha}$ for that of $u(t)$.

The generation of additional frequency components by nonlinear systems is the result of taking sums and differences of the input frequency components. The order of a nonlinearity directly corresponds to the number of frequencies that are summed together. For instance, let $U(k)$ have components at harmonics $k_1, k_2$ and $k_3$. Considering the negative of each of the harmonics,

$$-k_1, -k_2, -k_3, k_1, k_2, k_3$$

a quadratic nonlinearity will yield components at all possible combinations of $k_i + k_j$:

$$-k_1 + (-k_1), -k_1 + (-k_2), \ldots, k_3 + k_3$$

Similarly an $\alpha^{th}$ order nonlinearity will group $\alpha$ harmonics (with repetitions) from a set of $F$ frequencies and in the presence of a second input signal, the grouping will include frequencies of that signal. In general, for an $\alpha^{th}$ order nonlinearity of two inputs the nonlinear effect will generate harmonics at:

$$\sum_{i_1,\ldots,i_{\alpha-n}=-F}^{F} \; \sum_{j_1,\ldots,j_n=-F}^{F} k_{i_1} + \cdots + k_{i_{\alpha-n}} + h_{j_1} + \cdots + h_{j_n} \quad n = 0, 1 \ldots \alpha \tag{4}$$

In equation (4) $k_{-F} \ldots k_F$ and $h_{-F} \ldots h_F$, are the corresponding harmonics of two input signals with $k_{-i} \equiv -k_i$.

Computing the frequencies generated by nonlinearities of all orders up to $\alpha$ is unnecessary. It suffices to perform the computation for orders $\alpha$ and $\alpha - 1$ since these two orders will duplicate the frequencies generated by all the lower order nonlinearities. This is because any order greater than two can always be factorised to give a second degree factor, for example $u^3(t)v(t) = u^2(t)u(t)v(t)$. Since a second degree factor generates a zero harmonic, the sums and differences will replicate the harmonics generated by the other factor; so $u^3(t)v(t)$ will duplicate all the harmonics generated by $u(t)v(t)$. For a replication of the frequencies from orders $\alpha$ and $\alpha - 1$, nonlinear orders $\alpha + 2$ and $\alpha + 1$ are required, which by definition do not exist.

# 3 Methodology

## 3.1 Method I: Three stage experiment

A description of a system as given in equation (3) allows the nonlinear distortions to be grouped into contributions purely from each of the inputs and their combinations.

$$Y_d(k) = D_1(k) + D_2(k) + D_3(k) \tag{5}$$

In the above expression $D_1(k)$ is the distortion purely due to input $u(t)$ (the response to the first order kernel is not considered), $D_2(t)$ to that of $v(t)$ and $D_3(t)$ the distortions from combinations of both inputs. Grouping the

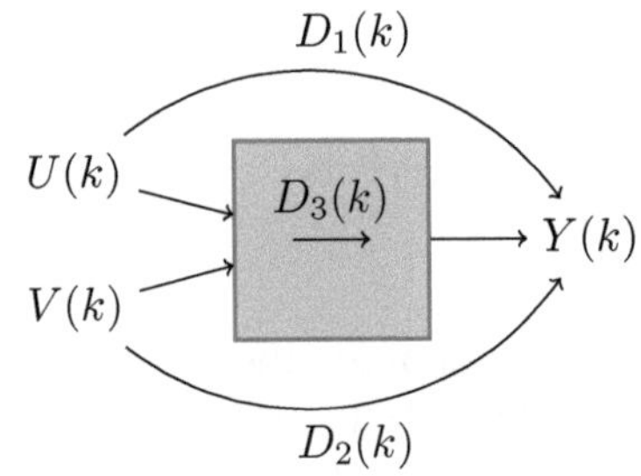


Figure 1: TISO: $D_1(k)$: Contributions from $u(t)$. $D_2(k)$: Contributions from $v(t)$. $D_3(k)$: Contributions from both inputs.

distortions to that of purely $U(k)$, $V(k)$ and their combinations gives:

$$D_1(k) = \sum_{\alpha=2}^{\infty} \sum_{l_1,\ldots,l_{\alpha-1}=-M/2}^{M/2} G_{\alpha U_\alpha V_0}(l_1,\ldots,l_{\alpha-1}) \prod_{i=1}^{\alpha} U(l_i) \tag{6}$$

$$D_2(k) = \sum_{\alpha=2}^{\infty} \sum_{p_1,\ldots,p_{\alpha-1}=-M/2}^{M/2} G_{\alpha U_0 V_\alpha}(p_1,\ldots,p_{\alpha-1}) \prod_{i=1}^{\alpha} V(p_i) \tag{7}$$

$$D_3(k) = \sum_{\alpha=2}^{\infty} \sum_{j=1}^{\alpha-1} \sum_{\substack{l_1,\ldots,l_{\alpha-j-1}=-M/2 \\ p_1,\ldots,p_{j-1}=-M/2}}^{M/2} G_{\alpha U_{\alpha-j} V_j}(l_1,\ldots,l_{\alpha-j-1},p_1,\ldots,p_{j-1}) \prod_{i=1}^{\alpha-j} U(l_i) \prod_{i=1}^{j} V(p_i) \tag{8}$$

From equations (6), (7) and (8) it readily follows that, if one of the inputs is set at zero the distortion at a frequency of interest in the output spectrum is purely a result of the other excited input. The individual nonlinear contributions and the contribution from their combinations can therefore be evaluated as follows:

1. Set input $v(t)$ at zero, yielding nonlinear distortions purely due to $u(t), D_1(k)$.
2. Set input $u(t)$ at zero, yielding nonlinear distortions purely due to $v(t), D_2(k)$.
3. Excite both inputs $u(t)$ and $v(t)$ giving rise to the total nonlinear distortions. For a frequency of interest deduct $D_1(k) + D_2(k)$ (obtained from the previous steps) from $Y(k)$ yielding the nonlinear distortions due to the input combinations, $D_3(k)$.

The practicality of setting each of the inputs at zero may be infeasible; therefore the procedure described is an ideal case. The alternative to setting an input at zero is to fix each of the inputs at its nominal operating point, while exciting the other input. The procedure will result in distortions due to both inputs. For example, let the operating point of input $v(t)$ be $c_v$ for all values of time. Since the signal $v(t)$ only has a component at zero frequency, let this be $V(0) = C_v$ equations (7) and (8) reduce to:

$$D_2(0) = \sum_{\alpha=2}^{\infty} \sum_{p_1,\ldots,p_{\alpha-1}=-M/2}^{M/2} C_v^{\alpha} G_{\alpha U_0 V_\alpha}(p_1 = 0,\ldots,p_{\alpha-1} = 0) \tag{9}$$

$$D_3(k) = \sum_{\alpha=2}^{\infty} \sum_{j=1}^{\alpha-1} \sum_{\substack{l_1,\ldots,l_{\alpha-j-1}=-M/2 \\ p_1,\ldots,p_{j-1}=-M/2}}^{M/2} C_v^{j} G_{\alpha U_{\alpha-j} V_j}(l_1,\ldots,l_{\alpha-j-1},p_1 = 0,\ldots,p_{j-1} = 0) \prod_{i=1}^{\alpha-j} U(l_i) \tag{10}$$

Equation (9) is simply the contributions of input $v(t)$ to the zero frequency and equation (10) shows that the nonlinear distortions due to both inputs simplify to distortions from input $u(t)$ conditioned by the operating point of signal $v(t)$. Therefore this term ($D_3(k)$) can be freely attributed to $D_1(k)$ as a distortion purely due to $u(t)$, provided that the operating level remains constant. As such the levels of distortions at the output amplitude spectrum, when the signals have a constant operating value can be written as:

With $v(t)$ set to its operating point:

$$Y_d(k) = D_1(k) + D_3(k) = D_1^a(k) \tag{11}$$

With $u(t)$ set to its operating point:

$$Y_d(k) = D_2(k) + D_3(k) = D_2^b(k) \tag{12}$$

With both inputs excited:

$$Y_d(k) = D_1^a(k) + D_2^b(k) + D_3^c(k) \tag{13}$$

The nonlinear distortion levels, $D_1^a, D_2^b$ and $D_3^c$ can be similarly determined as before:

1. Apply $u(t)$, set input $v(t)$ at its operating point, yielding nonlinear distortions due to $u(t), D_1^a(k)$.
2. Apply $v(t)$, set input $u(t)$ at its operating point, yielding nonlinear distortions due to $v(t), D_2^b(k)$.
3. Excite both inputs $u(t)$ and $v(t)$ giving rise to the total nonlinear distortions. For a frequency of interest deduct $D_1^a(k) + D_2^b(k)$ (obtained from the previous steps) from $Y(k)$ yielding the nonlinear distortions due to the input combinations, $D_3^c(k)$.

In the presence of output measurement noise, several periods of each input are required to obtain an estimate of $D_1$, $D_2$ or $D_3$ (similarly for $D_1^a$, $D_2^b$ or $D_3^c$). The DFT of each output period is averaged to minimise the effect of noise and allows the noise variance to be estimated as function of frequency. Further, as the phases of the multisine signals are randomly chosen, the estimates dependence on the signals phase must be considered. This is achieved by repeating the procedure with two newly generated multisine signals having new phase realisations. By repeating for several new multisine signals an average estimate can be obtained.

## 3.2 Method II: Detection via unique combinations

If the highest order of the system's nonlinearities is known, the frequencies at the output spectrum can be analytically evaluated by computing the sums and differences of all the input frequency components as given through equation (4). Similar to the concept in Method I for a given nonlinear order $\alpha$, the output frequencies can be grouped as those being a result purely from each input or their combinations.

$$S_u = \sum_{i_1,\ldots,i_{\alpha-n}=-F}^{F} k_{i_1} + \cdots + k_{i_{\alpha-n}} \qquad n = 0 \tag{14}$$

$$S_v = \sum_{j_1,\ldots,j_n=-F}^{F} h_{j_1} + \cdots + h_{j_n} \qquad n = \alpha \tag{15}$$

$$S_{uv} = \sum_{i_1,\ldots,i_{\alpha-n}=-F}^{F} \sum_{j_1,\ldots,j_n=-M/2}^{M/2} k_{i_1} + \cdots + k_{i_{\alpha-n}} + h_{j_1} + \cdots + h_{j_n} \qquad n = 1\ldots\alpha-1 \tag{16}$$

In equations (14-16), $S_u$ is the set of harmonics in the output generated by the frequencies of signal $u(t)$, $S_v$ from that of $v(t)$ and $S_{uv}$ from combinations of both input frequencies. If the input frequencies can be chosen such that there exists at least one unique harmonic in each set, the two signals can be used to classify the nonlinearities. By observing the output spectrum for the presence of any unique harmonic, this will indicate whether nonlinearities purely due to each input or their combinations are present.

The search for input frequencies that generate at least one unique harmonic in each of the sets, is carried out via a computational search. The algorithm begins by defining the largest allowed harmonic, $B$, for the two inputs and the number of desired harmonics, $F$, within this bandwidth. The interval is divided by the number of harmonics, creating as many subintervals as there are harmonics and randomly selecting an odd harmonic from each subinterval. The two input signals will therefore have a different harmonic specification and let the highest selected harmonic be $k_{max}$. With the highest nonlinear order defined as $\alpha$, the harmonic sets $S_u$, $S_v$ and $S_{uv}$ are computed for orders $\alpha$ and $\alpha - 1$. The corresponding input harmonic specifications will only be stored in memory if there exists at least one distinct harmonic in each set and is less than or equal to the highest input harmonic, $k_{max}$. The algorithm is repeated as many times desired, saving all successful harmonic specifications for the given value of $B$ and $F$.

In summary, the input harmonic specification is considered successful if none of the sets defined in equations (17-19) is empty:

$$SS_u = \{k_i | k_i \leq k_{max}, k_i \in S_u, k_i \notin S_v, k_i \notin S_{uv}\} \quad (17)$$
$$SS_v = \{k_i | k_i \leq k_{max}, k_i \notin S_u, k_i \in S_v, k_i \notin S_{uv}\} \quad (18)$$
$$SS_{uv} = \{k_i | k_i \leq k_{max}, k_i \notin S_u, k_i \notin S_v, k_i \in S_{uv}\} \quad (19)$$

As the number of harmonics in $SS_u$, $SS_v$ or $SS_{uv}$ may differ for different input harmonic configurations, a cost function is defined to evaluate the optimal input harmonic specification. Defining the cost function as $V$, it is the maximum-minimum of the number of harmonics in sets $SS_u$, $SS_v$ and $SS_{uv}$ with the optimal harmonic specification yielding the maximum value for $V$.

$$V = max\{min\{|SS_u|, |SS_v|, |SS_{uv}|\}\} \quad (20)$$

# 4 Experimental Results

## 4.1 Experimental results from Method I

Method I described earlier is applied to an electrical TISO system with the following schematic:

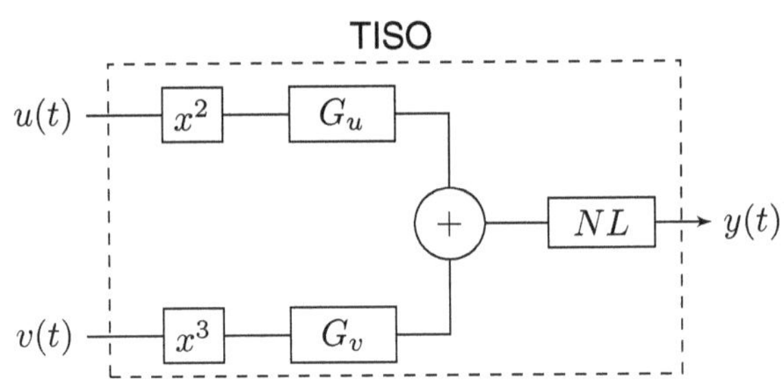


Figure 2: Experimental setup

Here $G_u$ and $G_v$ are two linear low pass filters, with cut-off frequency at approximately 6KHz. $NL$ is a nonlinear system with a bandwidth of approximately 1.6KHz. The two inputs $u(t)$ and $v(t)$ are multisines with seven randomly selected odd harmonics from a bandwidth of fundamental frequency $f_0 = 100$Hz and maximum frequency $f_{max} =$ 3KHz. The harmonic amplitudes are equal and selected to give each signal a root mean square (rms) value of 0.5V.

Two cases are presented here. Firstly the nonlinear distortion levels are estimated with operating points at 0V, secondly the operating point of input $u(t)$ is set at 1V and 2V for input $v(t)$.

**Case 1: Operating points at zero**

A single experiment involves the application of the signals in three stages as described in Method I. During each stage, four periods of the output signal and input signals are measured and averaged to minimise the noise effects. The nonlinear distortions are estimated over ten repeated experiments with multisine signals having newly generated phases but having the same harmonic specification as used in the first experiment.

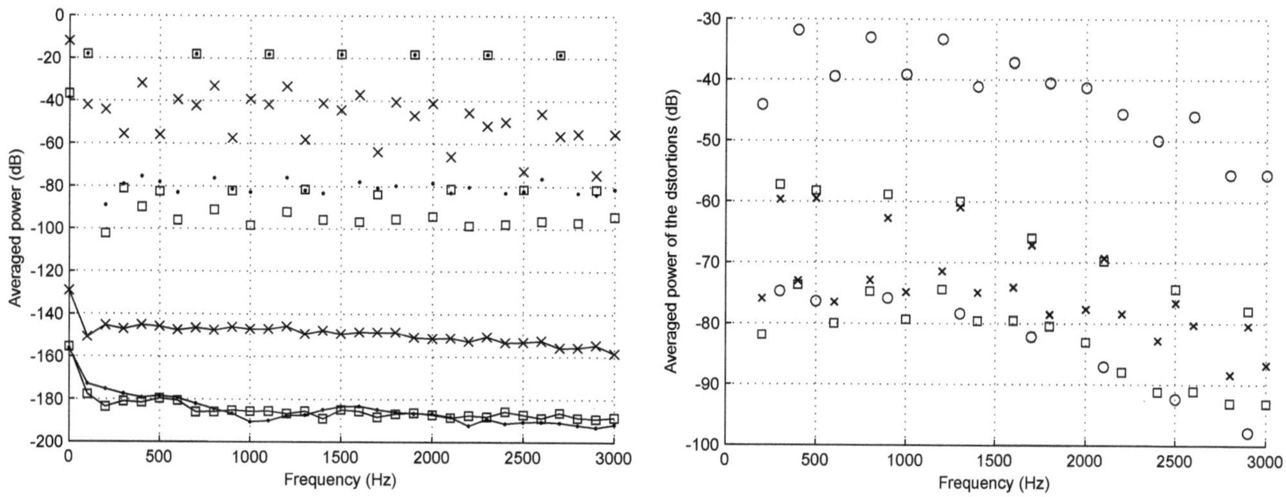


(a) Power spectrum of input and output signals. Dot: Signal $u(t)$; Square: $v(t)$; Cross: $y(t)$; Solid line with dots: Noise at $u(t)$; Solid line with squares: Noise at $v(t)$; Solid line with crosses: Noise at $y(t)$

(b) Power spectrum of nonlinear distortions. Circle: From $u(t)$, $D_1(k)$; Square: From $v(t)$, $D_2(k)$; Cross: From both, $D_3(k)$

Figure 3: Power spectra with zero operating points

Figure 3(a) shows the desired flat power spectrum in the seven randomly selected odd harmonics (approximately at -20dB) of the input signals. The crosses in the figure mark the averaged power spectrum of the output signal when both inputs are excited. The average noise power estimated for each input and the output is shown by the solid lines with markers. The noise levels are significantly lower than the signal powers, around -190dB at the inputs and about -150dB at the output. As such the power levels of the harmonics (not specified in the inputs) in the output spectrum can be distinguished as significant nonlinear distortions. The estimations of the nonlinear distortions from each input and their combinations are shown in Figure 3(b). Here, only the power levels at the non specified harmonics are shown. It is seen that distortions purely due to input $u(t)$ dominate, while the contributions purely due to $v(t)$ and their combinations are of equal magnitude. As such, it can be concluded that with these operating levels the contribution from input $u(t)$ to the output $y(t)$ is more nonlinear than from $v(t)$ to $y(t)$ or their combinations.

**Case 2: With nonzero operating points**

The operating points are now set at 1V for input $u(t)$ and 2V for input $v(t)$. Again 4 periods are measured to average out the measurement noise and then repeated over 10 experiments to estimate the nonlinear distortions.

The flat power spectrum of the input signals along with the output power spectrum are shown in Figure 4(a). Similar noise levels as for the zero operating point case are estimated at the inputs and the output. From Figure 4(b) it is evident that the highest nonlinear contributions to the output are from nonlinear combinations of both inputs, while those purely due to input $u(t)$ have a lesser magnitude and those purely due to $v(t)$ have the least. This experiment also illustrates the dependency of a nonlinear system on the characteristics of the driving inputs. With nonzero operating points, the nonlinearities which are functions of both inputs dominate, while with zero operating points those purely due to input $u(t)$ dominate.

## 4.2 Simulation results for Method II

Two orders of nonlinearities are considered. Firstly, with $\alpha = 3$ (all powers up to 3 are included) and $\alpha = 5$ (all powers up to 5 are included). The highest allowed harmonic for each signal is set at $B = 200$ and $F$ the number of odd harmonics in this bandwidth is increased from 1 to 25. After 1000 repeated trials the successful (if the sets in equations (17-19) are not empty) input harmonic configurations for the given value of $F$ are saved.

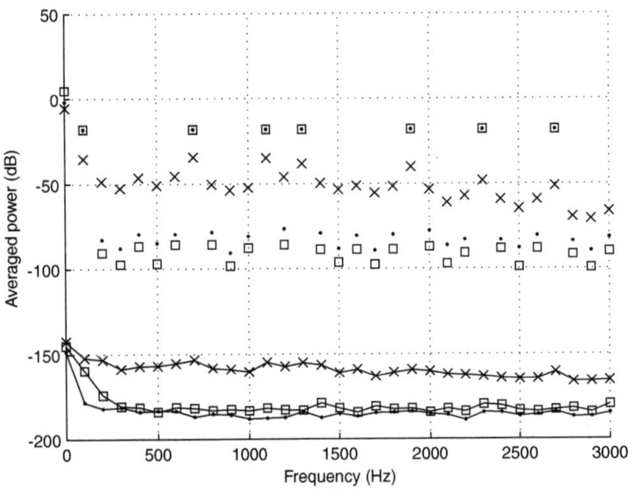


(a) Power spectrum of input and output signals. Dot: Signal $u(t)$; Square: $v(t)$; Cross: $y(t)$; Solid line with dots: Noise at $u(t)$; Solid line with squares: Noise at $v(t)$; Solid line with crosses: Noise at $y(t)$

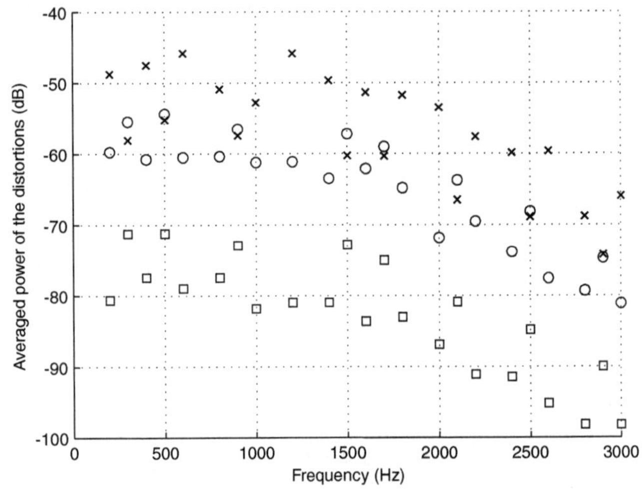


(b) Power spectrum of nonlinear distortions. Circle: From $u(t)$, $D_1^a(k)$; Square: From $v(t)$, $D_2^b(k)$; Cross: From both, $D_3^c(k)$

Figure 4: Power spectra with nonzero operating points

### 3rd order nonlinearity

Table 1 shows the number of successful (Successes) input harmonic configurations after 1000 repeated trails while increasing the number of odd harmonics in the specified bandwith.

| Odd harmonics $F$ | 2 | 3 | 4 | 5 | 6 | 7 | 8 | 9 | 10 |
|---|---|---|---|---|---|---|---|---|---|
| Successes | 918 | 983 | 995 | 998 | 999 | 999 | 999 | 995 | 989 |
| Odd harmonics | 11 | 12 | 13 | 14 | 15 | 16 | 17 | 18 | |
| Successes | 914 | 698 | 410 | 234 | 48 | 24 | 1 | 2 | |

Table 1: Number of odd harmonics and number of successful harmonic combinations

The results indicate that a single frequency (sinusoids) for each input and similarly inputs with more than 18 harmonics do not satisfy the conditions in equations (17-19) and fail to classify a cubic order nonlinearity. To evaluate which among the successful harmonic configurations is optimal, the cost function $V$ is evaluated and the results tabulated in Table 2:

| Odd harmonics $F$ | 2 | 3 | 4 | 5 | 6 | 7 | 8 | 9 | 10 |
|---|---|---|---|---|---|---|---|---|---|
| $V$ | 4 | 14 | 14 | 14 | 14 | 14 | 12 | 9 | 8 |
| Odd harmonics | 11 | 12 | 13 | 14 | 15 | 16 | 17 | 18 | |
| $V$ | 6 | 5 | 4 | 3 | 2 | 1 | 1 | 1 | |

Table 2: Cost function evaluated on all successful input harmonic specifications

Table 2 shows that signals with 3 to 7 randomly selected odd harmonics will yield at least 14 unique harmonics for the nonlinear classification. Further, as it is preferable to use signals with the most number of harmonics, signals having 7 randomly selected odd harmonics (in a bandwidth of $B = 200$) is the optimal choice. Table 5 shows the 7 odd harmonics for the two input signals that result in the highest cost function along with the unique harmonics in the output spectrum that classify the nonlinearities.

**5th order nonlinearity**

The simulation results with a fifth order nonlinearity are tabulated in the subsequent tables. Table 3 shows that as the number of odd harmonics is increased in the specified bandwidth, a single frequency component and using more 5 odd harmonics in the input signals fail to classify a fifth order nonlinearity.

| Odd harmonics $F$ | 2 | 3 | 4 | 5 |
|---|---|---|---|---|
| Successes | 518 | 134 | 6 | 1 |

Table 3: Number of odd harmonics and number of successful harmonic combinations

The outcome of the cost function evaluated on the successful harmonic configurations, given in Table 4, indicates that using 3 odd harmonics is the optimal choice, yielding at least 5 unique harmonics to classify the nonlinearities. The harmonics that result in the highest cost function along with the unique harmonics that classify the nonlinearities

| Odd harmonics $F$ | 2 | 3 | 4 | 5 |
|---|---|---|---|---|
| $V$ | 4 | 5 | 1 | 1 |

Table 4: Cost function evaluated on all the successful input harmonic specification

are given in Table 6.

comparing tables 1 and 3, it is seen that the number harmonics within a specified bandwidth, capable of classifying the nonlinearity, greatly decreases with the increase in nonlinear order. For systems with nonlinear order greater than 5, the method yields signals with harmonics much further apart making them impractical to use.

# 5 Conclusions

In this paper it is demonstrated that for a TISO system the nonlinear contributions to the output spectrum can be classified in relation to each of the inputs or their combinations. For a given set of operating points, Method I identifies the dominant nonlinear contribution. In this method the signals for the two inputs must have the same input harmonics and be retained for the application in the subsequent stages. Method II presents with an alternative mode of detecting and classifying the nonlinearity. As the order of the nonlinearity increases the number of input harmonics (via computational search) within a specified bandwidth satisfying the necessary conditions decreases greatly. As such for Method II, harmonics of input signals become more sparse with higher order nonlinearities rendering them impractical.

Experimental results on an electronic filter and using Method I illustrates that the dominant nonlinear contribution, with zero operating levels, is purely due to a single input and nonlinear combinations of both inputs are more dominant with nonzero operating points. The simulation results for Method II provides with candidate input harmonic specifications to detect and classify a cubic and fifth order nonlinearity. The results also demonstrate the sparseness of the input harmonics as the order is increased from a cubic to a fifth.

| For $u(t)$ $k$ | For $v(t)$ $k$ | From $u(t)$ $SS_u$ | From $v(t)$ $SS_v$ | Both $SS_{uv}$ |
|---|---|---|---|---|
| 1 | 21 | 18 | 10 | 4 |
| 33 | 31 | 24 | 40 | 8 |
| 69 | 61 | 34 | 52 | 12 |
| 87 | 91 | 36 | 82 | 14 |
| 137 | 129 | 50 | 84 | 22 |
| 161 | 149 | 74 | 98 | 28 |
| 195 | 175 | 86 | 112 | 43 |
| | | 102 | 114 | 48 |
| | | 120 | 122 | 56 |
| | | 126 | 144 | 64 |
| | | 136 | 152 | 76 |
| | | 138 | 154 | 80 |
| | | 156 | 180 | 90 |
| | | 194 | 190 | 94 |
| | | | | 96 |
| | | | | 100 |
| | | | | 106 |
| | | | | 116 |
| | | | | 124 |
| | | | | 130 |
| | | | | 134 |
| | | | | 140 |
| | | | | 142 |
| | | | | 147 |
| | | | | 148 |
| | | | | 158 |
| | | | | 164 |
| | | | | 168 |
| | | | | 176 |
| | | | | 178 |
| | | | | 192 |
| Total | | 14 | 14 | 31 |

Table 5: Cubic: Harmonic specification that results in the highest cost function

| For $u(t)$ $k$ | For $v(t)$ $k$ | From $u(t)$ $SS_u$ | From $v(t)$ $SS_v$ | Both $SS_{uv}$ | |
|---|---|---|---|---|---|
| 11 | 63 | 44 | 15 | 1 | 80 |
| 103 | 115 | 55 | 37 | 3 | 82 |
| 177 | 165 | 59 | 100 | 9 | 83 |
| | | 70 | 128 | 10 | 84 |
| | | 173 | 152 | 12 | 86 |
| | | | | 17 | 90 |
| | | | | 19 | 93 |
| | | | | 21 | 95 |
| | | | | 23 | 97 |
| | | | | 25 | 101 |
| | | | | 27 | 105 |
| | | | | 28 | 111 |
| | | | | 30 | 112 |
| | | | | 31 | 116 |
| | | | | 34 | 123 |
| | | | | 35 | 124 |
| | | | | 38 | 127 |
| | | | | 42 | 131 |
| | | | | 43 | 135 |
| | | | | 49 | 138 |
| | | | | 53 | 142 |
| | | | | 54 | 145 |
| | | | | 62 | 149 |
| | | | | 64 | 153 |
| | | | | 69 | 156 |
| | | | | 71 | 157 |
| | | | | 72 | 164 |
| | | | | 73 | 168 |
| | | | | 75 | 169 |
| | | | | 77 | 175 |
| | | | | 79 | |
| Total | | 5 | 5 | 61 | |

Table 6: Fifth order: Harmonic specification that results in the highest cost function

## References


[1] Evans, C., Rees, D., Jones, L., and Weiss, M., Periodic signals for measuring non-linear Volterra kernels, *IEEE Transactions on Instrumentation and Measurement* **45**, 362–371, 1996.

[2] Dobrowiecki, T. P. and Schoukens, J., Measuring a linear approximation to weakly nonlinear MIMO systems, *Automatica* **43**, 1737–1751, 2007.

[3] Dobrowiecki, T. P., Schoukens, J., and Guillaume, P., Optimized Excitation Signals for MIMO Frequency Response Function Measurements, *IEEE Transactions on Instrumentation and Measurement* **55**(6), 2072–2079, 2006.

[4] Godfrey, K., *Perturbation Signals for System Identification*, Prentice Hall, 1993.

[5] Pintelon, R. and Schoukens, J., *System Identification - A Frequency Domain Approach*, IEEE Press, 2001.

[6] Schetzen, M., *The Volterra and Wiener Theories of Nonlinear Systems*, New York: Wiley, 1980.

[7] Evans, C., Rees, D., and Jones, L., Non-linear disturbance errors in system identification using multisine test signals, *IEEE Transactions on Instrumentation and Measurement* **43**, 238–244, 1994.

[8] D'haene, T., Pintelon, R., Schoukens, J., and Van Gheem, E., Variance Analysis of Frequency Response Function Measurements Using Periodic Excitations, *IEEE Transactions on Instrumentation and Measurement* **54**(4), 1452–1456, 2005.

[9] Schoukens, J., Lataire, J., Pintelon, R., Vandersteen, G., and Dobrowiecki, T., Robustness Issues of the Best Linear Approximation of a Nonlinear System, *IEEE Transactions on Instrumentation and Measurement* **58**(5), 1737–1745, 2009.

Proceedings of the IMAC-XXVIII
February 1–4, 2010, Jacksonville, Florida USA


# A multi-harmonic approach to updating locally nonlinear structures

**I. Isasa[1], S. Cogan[2], E. Sadoulet-Reboul[2] and J.M. Abete[3]**

[1] Ikerlan - IK4, Department of Mechanical Engineering
J.M. Arizmendiarreta 2, 20500, Arrasate-Mondragon, Spain

[2] FEMTO-ST Institute, Structural Dynamics Research Group
Rue de l'Epitaphe 24, 25000 Besançon, France

[3] Mondragon Unibertsitatea, Department of Mechanical Engineering
Loramendi 4, 20500, Arrasate-Mondragon, Spain

e-mail: **iisasa@ikerlan.es**

**Abstract**

Improving the fidelity of numerical simulations using available test data is an important activity in the overall process of model verification and validation. While model updating or calibration of linear elastodynamic behaviors has been extensively studied for both academic and industrial applications over the past three decades, methodologies capable of treating nonlinear dynamics remain relatively immature. The authors propose a novel strategy for updating an important subclass of nonlinear models characterized by globally linear stiffness and damping terms in the presence of local nonlinear effects. The approach combines two well-known methods for structural dynamic analysis. The first is the Multi-harmonic Balance (MHB) method for solving the nonlinear equations of motion of a mechanical system under periodic excitation. This approach has the advantage of being much faster than time domain integration procedures while allowing a wide range of nonlinear effects to be taken into account. The second method is the Extended Constitutive Relation Error (ECRE) that has been used in the past for error localization and updating of linear elastodynamic models. The proposed updating strategy will be illustrated using academic examples.

## 1 Introduction

Nonlinear phenomena are commonplace in mechanical systems containing mechanisms, joints and contact interfaces [1]. Engineers often simplify the behavior of complex structural models by considering them to be linear for dynamic analyses, thus neglecting nonlinear effects due to large displacements, contact, clearance and impact phenomena, among others.

The following paper is devoted to the revision of nonlinear models in the field of structural dynamics based on measured responses. During the past two decades, linear model updating has been extensively studied to improve the accuracy of simulations [2]. Nonlinear model updating techniques on the other hand have received much less attention. Both time domain or frequency domain approaches can be found in the literature. In the time domain, the Restoring Force Surface method (RFS) and Proper Orthogonal Decomposition (POD) are described in detail in the overview paper by Kerschen et al. [3] with complete references to the literature. More recently, Gondhalekar et al. has proposed a strategy combining the RFS method with model reduction [4]. In the frequency domain, Böswald and Link [5] have developed a methodology based on the first order Harmonic Balanced method to obtain a suitable representation of nonlinear effects and they have applied their approach to update nonlinear joint parameters in complex structural assemblies. Another frequency domain method is investigated by Puel [6] where the Extended Constitutive Relation Error (ECRE) for linear dissipative systems is generalized to nonlinear model updating with a first order harmonic balance approximation. With the exception of the RFS method, the existing methods for nonlinear updating are based on some form of linearization and this naturally limits their application to relatively weak nonlinear effects. As for the RFS approach, its major weakness lies in the fact that it requires that the structural responses be measured on all model degrees-of-freedom where significant nonlinear effects are present.

In this paper, a novel methodology is presented which effectively combines the Multi-harmonic Balance method for calculating the periodic response of a nonlinear system and the Extended Constitutive Relation Error method for establishing a well-behaved metric for modeling and test-analysis errors. The proposed approach is not based on any linear approximations nor does it require the observation of all nonlinear degrees-of-freedom. An academic beam example with simulated experimental data will be used to illustrate the advantages and limitations of the methodology.

# 2 Mathematical formulation

## 2.1 Equations of motion

The equations of motion of a discrete linear structure can be written:

$$M\ddot{q}(t) + C\dot{q}(t) + Kq(t) = p(t) \tag{1}$$

where, $K, M, C \in \Re^{N,N}$ are respectively the symmetric stiffness, mass and damping matrices, with the stiffness matrix assumed to be non-negative definite; $p(t) \in \Re^{N,1}$ is a vector of external forces; $q(t) \in \Re^{N,1}$ is the vector of time responses on the $N$ degrees-of-freedom (dofs).

The equation of motion of a nonlinear structure can be written in the same way as a linear structure with the addition of a nonlinear term, $f_{NL}(q(t), \dot{q}(t)) \in \Re^{N,1}$, which can depend on the system displacements and velocities:

$$M\ddot{q}(t) + C\dot{q}(t) + Kq(t) + f_{NL}(q(t), \dot{q}(t)) = p(t) \tag{2}$$

The origin of these nonlinear forces can be quite diverse, including:

- Some large displacement systems, for example the classical pendulum.
- Material nonlinearities including locally plastic or viscoplastic behaviors, shape memory alloys, and so on.
- Local interface nonlinearities including Hertz contact, dry friction, intermittent contact or clearance phenomena.

The response of a nonlinear system can be qualitatively very different from a linear one. In a linear system the steady-state response to a periodic excitation is at the same frequency as the excitation force once the transient term vanishes in time and is independent of the initial conditions. The periodic response of a nonlinear system, when it exists, generally exhibits primary and secondary resonances and may depend on the initial conditions [7].

Although transient behavior may be important, the study of periodic solutions and their stability remains essential to capturing the behavior of a vibrating system. Nonlinear time domain simulations are extremely burdensome especially when they are used to calculate the steady state response of large-order models. The Multi-harmonic Balance method is based on a Fourier series approximation and was developed with the objective of solving for the periodic response of nonlinear systems more efficiently.

## 2.2 Multi-harmonic balance method

The MHB method is a frequency domain approach developed to solve equation (2) for a periodic excitation. Many extensions to the first-order harmonic balance approach to include higher harmonics were developed in the 1980's, for example [8] or [9]. We have based our developments on the formulation proposed by Cardona et al. [10] and more recently applied to complex industrial structures with contact effects by Petrov et al., for example [11].

The equilibrium equation of a nonlinear system of $N$ degrees-of-freedom is given by equation (2). Expressing the vector of time responses $q(t)$ as a Fourier series yields:

$$q(t) = Q_0 + \sum_{j=1}^{n} \left( Q_j^c \cos m_j \omega t + Q_j^s \sin m_j \omega t \right) \tag{3}$$

where:

- $Q_0$ represents the constant or static contribution
- $Q_j^c$ and $Q_j^s$ are respectively the $j^{th}$ cosine and sine coefficients of the Fourier series
- $m_j$ expresses the harmonic of the excitation frequency $\omega$

Introducing this expression into equation (2) yields:

$$\begin{gathered} K \left( Q_0 + \sum_{j=1}^{n} Q_j^c \cos m_j \omega t + Q_j^s \sin m_j \omega t \right) + \\ C \left( \sum_{j=1}^{n} -m_j \omega \, Q_j^c \sin m_j \omega t + m_j \omega \, Q_j^s \cos m_j \omega t \right) + \\ M \left( \sum_{j=1}^{n} -(m_j \omega)^2 \, Q_j^c \cos m_j \omega t - (m_j \omega)^2 \, Q_j^s \sin m_j \omega t \right) + \\ f(q(t), \dot{q}(t)) - p(t) = 0 \end{gathered} \tag{4}$$

A Galerkin procedure is then applied by sequentially pre-multiplying equation (4) by the harmonic functions $(1, \cos \omega t, \sin \omega t, \cos m_1 \omega t, \ldots, \cos m_n \omega t, \sin m_n \omega t)$ and integrating over the period $T = 2\pi/\omega$.

Regrouping the resulting equations for each harmonic in the Fourier expansion, the following frequency domain expression can be obtained:

$$\mathcal{Z}(\omega)\mathcal{Q} + \mathcal{F}(\mathcal{Q}) - \mathcal{P} - 0 \tag{5}$$

where $\mathcal{Q} = \{Q_0 \; ; \; Q_1 \; ; \; Q_2 \; ; \; \cdots \; ; \; Q_{2n-1} \; ; \; Q_{2n}\}$ is the vector of harmonic coefficients with $Q_i \in \Re^{N,1}$. The matrix $\mathcal{Z} \in \Re^{(2n+1)N,(2n+1)N}$ is given by:

$$\mathcal{Z} = \\ \begin{bmatrix} K & 0 & 0 & \cdots & 0 & 0 \\ 0 & K-(m_1\omega)^2 M & m_1 \omega C & \cdots & 0 & 0 \\ 0 & -m_1 \omega C & K-(m_1\omega)^2 M & \cdots & 0 & 0 \\ \cdots & \cdots & \cdots & \cdots & \cdots & \cdots \\ 0 & 0 & 0 & \cdots & K-(m_n\omega)^2 M & m_n \omega C \\ 0 & 0 & 0 & \cdots & -m_n \omega C & K-(m_n\omega)^2 M \end{bmatrix} \tag{6}$$

and the vectors $\mathcal{F}; \mathcal{P} \in \Re^{(2n+1)N,1}$ corresponding respectively to the nonlinear forces and the external excitations are given by:

$$
\mathcal{F} = \left\{ \begin{array}{c} \int_0^T f_{NL}(q(t), \dot{q}(t))\, dt \\ \frac{\omega}{\pi} \int_0^T f_{NL}(q(t), \dot{q}(t)) \cos \omega t\, dt \\ \frac{\omega}{\pi} \int_0^T f_{NL}(q(t), \dot{q}(t)) \sin \omega t\, dt \\ \vdots \\ \frac{\omega}{\pi} \int_0^T f_{NL}(q(t), \dot{q}(t)) \cos m_n \omega t\, dt \\ \frac{\omega}{\pi} \int_0^T f_{NL}(q(t), \dot{q}(t)) \sin m_n \omega t\, dt \end{array} \right\} \tag{7}
$$

and

$$
\mathcal{P} = \left\{ \begin{array}{c} \int_0^T p(t)\, dt \\ \frac{\omega}{\pi} \int_0^T p(t) \cos \omega t\, dt \\ \frac{\omega}{\pi} \int_0^T p(t) \sin \omega t\, dt \\ \vdots \\ \frac{\omega}{\pi} \int_0^T p(t) \cos m_n \omega t\, dt \\ \frac{\omega}{\pi} \int_0^T p(t) \sin m_n \omega t\, dt \end{array} \right\} \tag{8}
$$

**Remarks**

- Equations (7) and (8) demonstration the time-frequency character of the MHB algorithm where it is generally much easier to evaluate the forces in the time domain and then transform back into the frequency domain.
- Equation (5) is generally solved using a predictor-corrector continuation scheme in order to follow the distorsions of the corresponding frequency responses [12].
- Model reduction can be used effectively for the linear system matrices in order to reduce the computational burden for very large models.

## 2.3 Extended Constitutive Relation Error

The Constitutive Relation Error was initially proposed by Ladavèze et al. in the early 1980s as an error estimator for finite element models [13]. An extended version for use in model updating was described in the early 1990s [14] taking into account both modeling error and test-analysis errors for linear elastodynamic behaviors. A discrete formulation of the approach for dissipative linear structures can be found in [15]. The basic philosophy of the ECRE methodology consists in dividing the relations of interest (constitutive behavior laws, equations of motion, measured displacements, initial conditions, etc.) into two groups: the reliable and the less reliable quantities. The solution to the problem is sought so as to satisfy the reliable equations exactly while minimizing the errors in the less reliable equations. The present paper will be restricted to nonlinear elastodynamic systems which contain only nonlinear stiffness errors. Extensions to nonlinear dissipative effects as well as combined errors in both linear and nonlinear properties can be formulated in an analogous manner.

Let $\mathcal{Q}_\omega$ and $\mathcal{V}_\omega$ be two admissible displacement fields of equation (5) and $D^2(\mathcal{Q}_\omega, \mathcal{V}_\omega)$ a measure of distance between the two vectors such that:

$$D^2(\mathcal{Q}_\omega, \mathcal{V}_\omega) = \|\mathcal{Q}_\omega - \mathcal{V}_\omega\|^2_{\mathcal{K}} \equiv (\mathcal{Q}_\omega - \mathcal{V}_\omega)^T \mathcal{K}(\mathcal{Q}_\omega - \mathcal{V}_\omega) \tag{9}$$

where, $\mathcal{K} \in \Re(2n+1)N, (2n+1)N$ is the multi-harmonic stiffness matrix corresponding to the linear system defined by:

$$\mathcal{K} = \begin{bmatrix} K & 0 & 0 & \cdots & 0 & 0 \\ 0 & K & 0 & \cdots & 0 & 0 \\ 0 & 0 & K & \cdots & 0 & 0 \\ \cdots & \cdots & \cdots & \cdots & \cdots & \cdots \\ 0 & 0 & 0 & \cdots & K & 0 \\ 0 & 0 & 0 & \cdots & 0 & K \end{bmatrix} \tag{10}$$

A multi-harmonic ECRE can be defined for nonlinear stiffness errors in the following way:

$$E^2_\omega = r^T_\omega \mathcal{K} r_\omega + \alpha \left(H\mathcal{Q}_\omega - \mathcal{Q}^e_\omega\right)^T \mathcal{K}_R \left(H\mathcal{Q}_\omega - \mathcal{Q}^e_\omega\right) \tag{11}$$

where,

- $r_\omega \equiv \mathcal{Q}_\omega - \mathcal{V}_\omega$, with $\mathcal{Q}_\omega \in \Re^{(2n+1)N,1}$ and $\mathcal{V}_\omega \in \Re^{(2n+1)N,1}$ two admissible displacement fields for multi-harmonic equation of motion equation (5).
- $\mathcal{Q}^e_\omega \in \Re^{(2n+1)n_e,1}$ is the vector of identified harmonic coefficients on the $n_e$ measurement degrees-of-freedom.
- $H \in \Re^{(2n+1)n_e,(2n+1)N}$ is a projection matrix allowing the model responses $\mathcal{Q}_\omega$ to be projected onto the set of $n_e$ measurement directions so as to account for the limited number of measurement dofs and any differences in local reference frames between the FE model and the experimental model.
- $\mathcal{K}_R \in \Re^{(2n+1)n_e,(2n+1)n_e}$ is the multi-harmonic stiffness matrix of the linear system reduced to the measurement degrees-of-freedom. In practice, the Guyan stiffness matrix is generally used.
- $\alpha$ is a real positive scalar allowing the relative confidence in the identified harmonic coefficients to be taken into account.

Equation (11) is composed of two terms. The first term is a measure of the modeling error whereas the second term is a measure of the distance between the experimentally identified harmonic coefficients and those predicted by the model. Both of these terms correspond to the less reliable quantities in the present ECRE formulation. The reliable quantities correspond to the equilibrium equations of the system expressed by equation (5).

Therefore, the minimization problem to be solved in this case is given by:

$$\begin{cases} \text{Minimize } E^2_\omega & = r^T_\omega \mathcal{K} r_\omega + \alpha \left\|H\mathcal{Q}_\omega - \mathcal{Q}^e_\omega\right\|^2_{\mathcal{K}_R} \\ \text{Under the constraint } \mathcal{K} r_\omega & = \mathcal{Z}(\omega)\mathcal{Q}_\omega + \mathcal{F} - \mathcal{P} \end{cases} \tag{12}$$

or again:

$$\min g = r^T_\omega \mathcal{K} r_\omega + \alpha(H\mathcal{Q}_\omega - \mathcal{Q}^e_\omega)^T \mathcal{K}_R (H\mathcal{Q}_\omega - \mathcal{Q}^e_\omega) + \gamma^T(\mathcal{K} r_\omega - Z(\omega)\mathcal{Q}_\omega - \mathcal{F} + \mathcal{P}) \tag{13}$$

where $g$ is the objective function and $\gamma \in \Re^{(2n+1)N,1}$ is a vector of Lagrange multipliers.

The stationarity conditions require:

$$\begin{aligned} \frac{\partial g}{\partial r_\omega} = 0 \quad &\Rightarrow \quad \mathcal{K}(2r_\omega + \gamma) = 0 \\ \frac{\partial g}{\partial Q_\omega} = 0 \quad &\Rightarrow \quad -\mathcal{Z}(\omega)\gamma - \frac{\partial \mathcal{F}}{\partial \mathcal{Q}_\omega}\gamma + 2\alpha H^T \mathcal{K}_R (H\mathcal{Q}_\omega - \mathcal{Q}_\omega^e) = 0 \\ \frac{\partial g}{\partial \gamma} = 0 \quad &\Rightarrow \quad \mathcal{K}r_\omega - \mathcal{Z}(\omega)\mathcal{Q}_\omega - \mathcal{F} + \mathcal{P} = 0 \end{aligned} \tag{14}$$

Eliminating $\gamma$ and regrouping the equations yields the following nonlinear matrix equation:

$$\begin{bmatrix} \mathcal{Z}(\omega) + \dfrac{\partial \mathcal{F}}{\partial \mathcal{Q}_\omega} & \alpha H^T \mathcal{K}_R H \\ \mathcal{K} & -Z(\omega) \end{bmatrix} \begin{Bmatrix} r_\omega \\ \mathcal{Q}_\omega \end{Bmatrix} + \begin{Bmatrix} 0 \\ -\mathcal{F} \end{Bmatrix} = \begin{Bmatrix} \alpha H^T \mathcal{K}_R \mathcal{Q}_\omega^e \\ -\mathcal{P} \end{Bmatrix} \tag{15}$$

**Remarks**

- Equation (15) requires the solution of a nonlinear system of order $2N(2n+1)$. It can be solved with a classical Newton-Raphson iterative procedure.
- The solution of equation (15) comprises two unknown vectors. First, the residual vector $r_\omega$ represents the displacement field resulting from the unbalanced forces in the multi-harmonic equations of motion and provides the basis for calculating the modeling error. Second, the response vector $\mathcal{Q}_\omega$ represents the experimental multi-harmonic response expanded onto all model dofs and provides a means for evaluating the test-analysis distances.
- Given the vectors $r_\omega$ and $\mathcal{Q}_\omega$, the total MBH-ECRE error equation (11) for the point in model space defined by the nominal linear system matrices and the nominal nonlinear model used to estimate the multi-harmonic nonlinear forces can now be evaluated.
- The model updating problem simply consists in minimizing the total MBH-ECRE error over the space defined by coefficients of the nonlinear model.
- $\mathcal{Q}_\omega^e$ represents the vector of experimentally identified harmonic coefficients. It is obtained directly from the experimentally observed time responses via the Fast Fourier Transform [16]:

$$\mathcal{Q}_j^c = \frac{1}{N_p} \sum_{k=0}^{N_p-1} q(k) \cos\left(\frac{2\pi}{N_p}kj\right) \tag{16}$$

$$\mathcal{Q}_j^s = \frac{-i}{N_p} \sum_{k=0}^{N_p-1} q(k) \sin\left(\frac{2\pi}{N_p}kj\right) \tag{17}$$

  where, $N_p$ is the number of points per period and $i$ is the imaginary number.
- Model reduction can again be effectively used here to minimize memory requirements and calculation times.

# 3 Illustrative academic example

The proposed methodology will be illustrated on a simulated academic example based on the COST action F3 project benchmark [17]. The model consists in a clamped linear beam attached to a thinner beam at one end. The main beam has a length of 0.7 m and a thickness of 0.014 m, whereas the thin beam has a length of 0.04 m with a thickness of 0.0005 m. Both beams have a width of 0.014 m and the material of both of them is steel with a Young's modulus of 2.11E11 MPa and a Poisson ratio of 0.3.

The structure is excited at node number 3 (see Figure 1) with a stepped sine excitation having an amplitude of 2. This amplitude level was chosen based on the results of [3] in order to insure a large enough deflection for nonlinear effects to come into play.

As stated in [3] and [18], the nonlinear behavior appears mainly in the first mode (30.76 Hz). A grounded cubic stiffness is introduced at node number 8 with the goal of modeling these nonlinearities. Moreover, in this example the influence of this cubic nonlinearity is studied only for the first mode and for the first harmonic. The nominal value of the nonlinear coefficient was chosen to be $6.1\ 10^9$ N/m [18].

The FRF is calculated between the excitation point (node 3) and the response point (node 8) and plotted in Figure 2 in order to visualize the distortion resulting from the nonlinear effects.

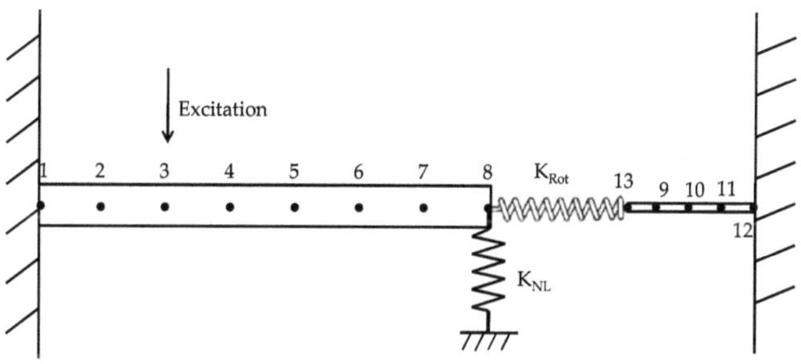


Figure 1: CostF3 beam.

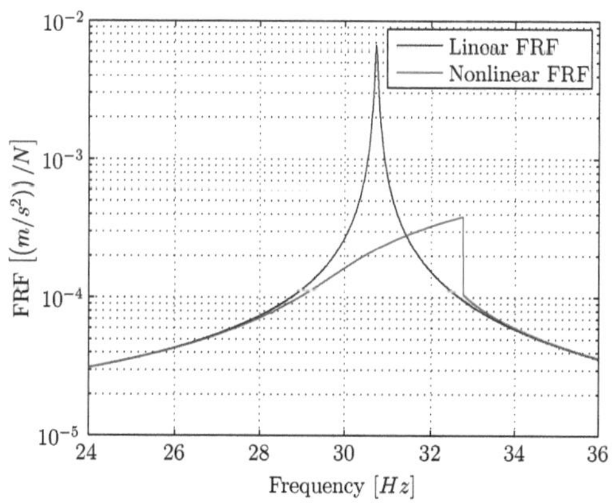


Figure 2: FRF between the excitation point (node 3) and the response point (node 8).

In order to apply the nonlinear model updating procedure based on the MBH-ECRE approach the values of the experimental vector of harmonic coefficients contained in $\mathcal{Q}^e_\omega$ are needed. In this paper they are obtained numerically using a Newmark algorithm developed based on [19] followed by a FFT analysis.

In order to illustrate the advantages and limitations of the proposed nonlinear updating strategy, it will be applied in three different simulated test configurations. In all three configurations, three excitation frequencies will be investigated corresponding to different response levels and thus different degrees of nonlinearity. The objective here is simply to examine the shape of the error expressed by equation (11) as a function of a single nonlinear model parameter. To simplify the interpretation of the results, the experimental harmonic coefficients have been generated based on the nominal nonlinear model. As such, in what follows we expect to see a minimum in the MBH-ECRE curve at a value of the correction coefficient that multiplies the nonlinear stiffness ($K_{NL}$) equal to 1.

- Test case 1

  The objective of this first configuration is simply to verify the implemented MBH-ECRE algorithm. When all model dofs have been measured, that is to say, all 21 dofs (10 translations and 11 rotations) of the beam in Figure 1 are observed, we obtain the exact results, as seen in Figure 3(b) for the three different excitation frequencies. It is noted that all three curves are convex and, as expected, are minimum for a correction coefficient equal to 1.

- Test case 2

  The second case aims at illustrating the impact of observing only a subset of a model dofs. In the present case, only 4 translations are assumed to be measured corresponding to nodes 3, 4, 6 and 8. Moreover, a model reduction has been performed based on the static Guyan procedure [20]. Finally, the model dof corresponding to the nonlinear cubic spring is assumed to be included in the set of observed dofs. Figure 3(c) shows the results of the MBH-ECRE for the three different excitation frequencies and once again it is noted that the curves are convex. However, the minimums of the curves are now shifted although the highest excitation frequency (corresponding to the highest response amplitude and thus the largest nonlinear effect) still has a minimum at very nearly 1. These shifts are due to the fact that the residual energies of the nonlinear spring are not taken into account in equation (11). However, as the nonlinear effects increase in magnitude, this discrepancy becomes less and less important. Future work will explicitly include these nonlinear residual energies.

- Test case 3

  This final case aims at illustrating a very important characteristic of the proposed updating strategy, namely that it is not necessary to experimentally observe the model degrees-of-freedom corresponding to the location of the nonlinear physics (the translational displacement at node 8 in this example). We retain the same reduced system matrices as in the previous test case but it is assumed that the displacement of node 8 is no longer available. The results are plotted in Figure 3(d) and are qualitatively similar to those of the previous test case.

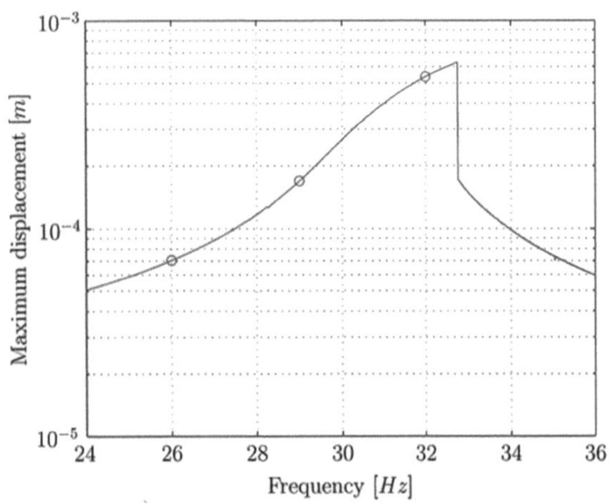


(a) Maximum displacement in node 8

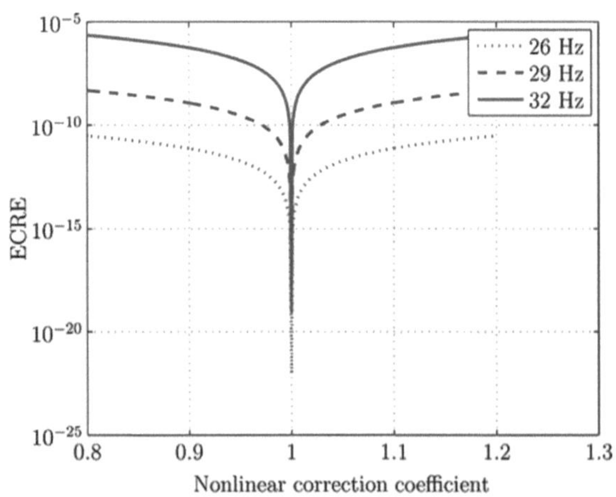


(b) ECRE when 21 dof are measured

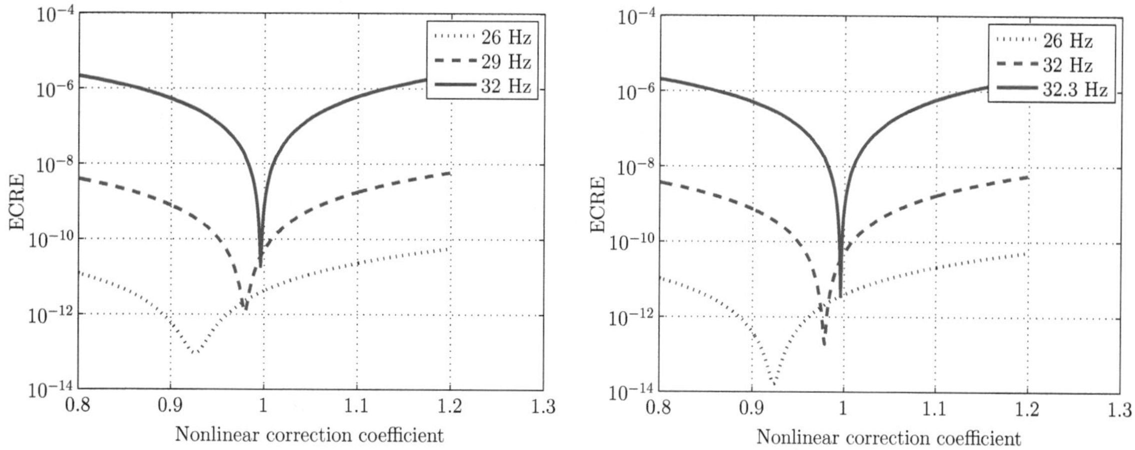


(c) ECRE when a reduced model of 4 dof is used and when only 4 dof are measured

(d) ECRE when a reduced model of 4 dof is used and when only 3 dof are measured, the nonlinear dof is not measured

Figure 3: Extended MHB-ECRE nonlinear model updating procedure on a beam example.

# 4 Conclusions

This paper presents a novel nonlinear model updating approach that combines two well-known strategies for structural dynamic analysis, namely the Multi-harmonic Balance method for calculating the periodic response of a nonlinear system and the Extended Constitutive Relation Error method for establishing a well-behaved metric for modeling and test-analysis errors. The proposed updating strategy has been illustrated using simulated data based on the COST-F3 beam benchmark. The potential advantages of this methodology are:

- The absence of a locally linearized model thus allowing strongly nonlinear systems to be addressed with the inclusion of higher-order harmonics.
- It is not necessary to experimentally observe the structural displacements at the location of the nonlinear physics.
- The model responses do not need to be re-evaluated at every updating iteration thus reducing the computational burden of the updating process.
- Model reduction techniques can be used very effectively to reduce calculation costs.

The main limitation of the MBH-ECRE method concerns the impact of measurement noise and harmonic truncation effects on the total MBH-ECRE curves. A decision indicator is currently under investigation to quantify the level of nonlinearity that can reasonably be identified for a given level of model reduction and measurement uncertainty.

# Acknowledgements

The work presented in this paper has been carried out with the generous support of Orona EIC in Spain.

# References

[1] Nayfeh, A. H. and Balachandran, B., *Applied Nonlinear Dynamics: Analytical, Computational and Experimental Methods*, Wiley Series in Nonlinear science, NY, USA, 1995.

[2] Mottershead, J.E. and Friswell, M.I., *Model Updating In Structural Dynamics: A Survey*, Journal of Sound and Vibration, Vol. 167, No. 2, pp. 347-375, 1993.

[3] Kerschen, G., Worden, K., Vakakis, A. F. and Golinval, J.C., *Past, present and future of nonlinear system identification in structural dynamics*, Mechanical Systems and Signal Processing, Vol. 20, No. 3, pp. 505-592, 2006.

[4] Gondhalekar, A. C., Petrov, E. P., Imregun, M., *Parameter identification in nonlinear dynamic systems via genetic algorithm optimization*, Journal of Computational Nonlinear Dynamics, Vol. 4, Issue 4, doi:10.1115/1.3187213, 2009.

[5] Böswald, M. and Link, M., *Identification of non-linear joint parameters by using frequency response residuals*, Proc. of the International Modal Analysis Conference IMAC-XXIII, Orlando, Florida, 2005.

[6] Puel, G., *Mise en évidence et recalage des non-linéarités locales en dynamique des structures*, Master's thesis, Université Paris VI, France, 2001.

[7] Worden, K and Tomlinson, G. R., *Nonlinearity in Structural Dynamics: Detection, Identification, and Modeling*, Institute of Physics Publishing (IoP), Bristol and Philadelphia, 2001.

[8] Pierre, C., Ferri, A., and Dowell, E. H., *Multi-harmonic analysis of dry friction damped systems using an incremental harmonic balance method*, ASME Journal of Applied Mechanics, Vol. 52, No. 4, 958-964, 1985.

[9] Cameron, T.M. and Griffin, J.H., *An alternating frequency / time domain method for calculating the steady-state response of nonlinear dynamic systems* ASME Journal of Applied Mechanics, VOl. 56, pp. 149-154, 1989.

[10] Cardona, A., Coune, T., Lerusse, A. and Geradin, M., *A multiharmonic method for non-linear vibration analysis*, International Journal for Numerical Methods in Engineering, Vol. 37, pp. 1593-1608, 1994.

[11] Petrov, E.P. and Ewins, D.J., *Effects of Damping and Varying Contact Area at Blade-Disk Joints in Forced Response Analysis of Bladed Disk Assemblies* Journal of Turbomachinery, Vol. 128, pp. 403-410, 2006.

[12] Ferreira, J.V. and Serpa, A.L., *Application of the arc-length method in nonlinear frequency response*, Journal of Sound and Vibration, Vol. 284, pp. 133-149, 2005.

[13] Ladevèze, P. and Leguillon, D., *Error estimate procedure in the finite element method and application*, SIAM Journal of Numerical Analysis, Vol. 20, No. 3, pp. 485-509, 1983.

[14] Ladevèze, P., Reynier, M. and Maia, N.M., *Error on the constitutive relation in dynamics: Theory and application for model updating*, Inverse Problems in Engineering, Balkema, Rotterdam, pp. 251-256, 1994.

[15] Barthe, D., Deraemaeker, A, Ladevèze, P. and Le Loch, S., *Validation and updating of industrial models based on the constitutive relation error*, AIAA Journal, Vol. 42, pp. 1427-1434, 2004.

[16] Berthillier, M., Dupont, C., Mondal, R. and Barrau, J.J., *Blades forced response analysis with friction dampers*, ASME Journal of Vibration and Acoustics, Vol. 120, pp. 468-474, 1998.

[17] Thouverez, F., *Presentation of the ECL Benchmark*, Mechanical Systems and Signal Processing, Vol. 17, No. 1, pp. 195-202, 2003.

[18] Lenaerts, V.,Kerschen, G. and Golinval, J.C., *ECL Benchmark: Application of the proper orthogonal decomposition*, Mechanical Systems and Signal Processing, Vol. 17, No. 1, pp. 237-242, 2003.

[19] Géradin, M. and Rixen, D., *Mechanical Vibrations : Theory and Applications to Structural Dynamics*, 2nd edition, Wiley series, 1997.

[20] Guyan, R.J., *Reduction of stiffness and mass matrices*, AIAA Journal, Vol. 3, No. 2, pp. 380-380, 1965.

Proceedings of the IMAC-XXVIII
February 1–4, 2010, Jacksonville, Florida USA


# A Block Rocking on a Seesawing Foundation

**L.N. Virgin**

Dept. Mechanical Engineering
Duke University
Durham, NC 27708-0300.

**Abstract**

Examples of vibro-impact mechanical and structural systems are not uncommon. Such systems are capable of exhibiting an interesting spectrum of non-smooth dynamic behavior when a characteristic changes abruptly. However, they present strong challenges to the analyst and designer trying to predict dynamic behavior. This paper considers an example of a rigid body (a rectangular block) that is placed on a flat surface that then oscillates in a seesawing, or tilting, motion. Given this kind of harmonic base excitation the free-standing block then responds accordingly, often by rocking. In this study the overturning characteristics of the block are of special interest. The motion is considered to take place in a vertical plane with no bouncing or sliding allowed. The outcome of experimental testing is compared with intensive numerical simulation.

## 1 INTRODUCTION

A rectangular block sits at rest on a flat rigid surface. The surface then starts to oscillate in a 'seesawing' motion, i.e., the angle of the surface is a harmonic function of time. A schematic of such a system is shown in Fig. 1, including the example of a container sitting on a floating deck in a sea-state. This is a relatively simple example of a strongly nonlinear system [1]. Another practical setting for this type of situation is the dynamic response of objects subject to an earthquake ground motion [2]. It has been shown that earthquakes can have a significant rotational component [3, 4].

We are specifically interested in the circumstances in which the block might overturn. Clearly, the more slender the block (in terms of an aspect ratio $H/B$) the more likely the block is to topple over. However, given a fixed geometry it is then natural to ask the question: what combination of forcing parameters cause overturning.

In this study we consider the rocking motion of a planar block. It is assumed that one corner of the block remains in contact with the base at all times, and that the block does not slide. Hence there is a single mechanical degree of freedom, the instantaneous angle, $\theta$, but a key point is this angle undergoes a sudden change as the block impacts and the sign of the angle changes. This type of piecewise linear characteristic underlies much of the complicated (nonlinear) dynamics to be described [5]. If the base is stationary then it is clear that overturning occurs when this angle exceeds the value $\tan^{-1}(H/B)$, i.e., when the center of mass, $C$, passes vertically over one of the corners (assuming the block is made of a homogeneous material). If the base oscillates with an angular motion $\alpha_0 = \psi \sin(\omega t)$ then the *impact* occurs when the relative angle (between the block and the base) reaches zero [6]. With forcing, the likelihood of overturning depends quite sensitively on the nature of the forcing. For example, if the magnitude of the forcing is large then the block is more susceptible to overturning, whereas if the base is oscillating very slowly and with a small magnitude then the block most likely wouldn't move (relative to the base) at all. But the dependence of overturning on the forcing parameters in the intermediate range is not so clear. In fact previous studies, based on extensive numerical simulation, have shown that the boundary in parameter space between overturning and not overturning is far from simple, and can even have fractal characteristics [7–9].

Previous work has focused on the numerical simulation of the nonlinear equation of motion. The present paper describes some

(a)

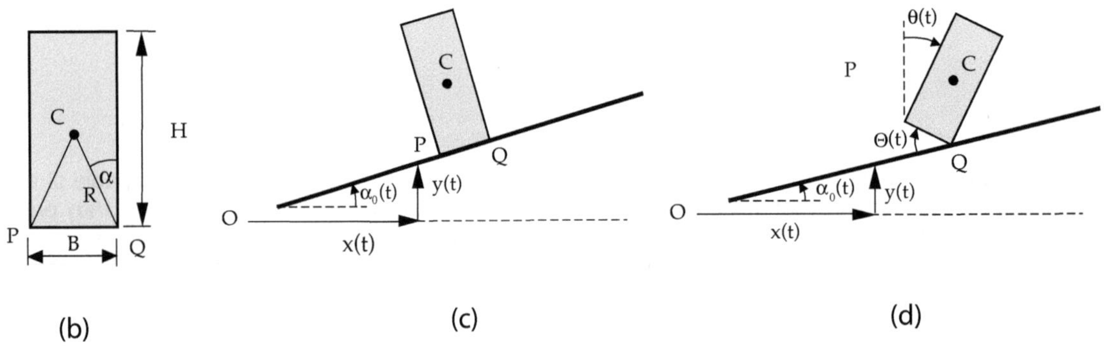


**Figure 1:** *(a) Schematic and (b) - (c) geometry of a rigid block rocking on a rolling deck.*

initial experimental results on this type of system.

## 2 THE NUMERICAL MODEL

Given the system shown in Fig. 1(b) - (d), and assuming the tilting of the base is harmonic, i.e., $\alpha_0(t) = \psi \sin \omega t$ we can write down a nondimensional equation of motion [7]:

$$\begin{aligned} &\theta'' - D\psi\Omega^2 \sin\alpha \sin\Omega\tau \sin(\alpha_0 - S\theta - S\psi\sin\Omega\tau) \\ &-SD\psi^2\Omega^2 \sin\alpha \cos^2\Omega\tau \cos(\alpha_0 - S\theta - S\psi\sin\Omega\tau) \\ &+S\sin(\alpha_0 - S\theta) = 0. \end{aligned} \tag{1}$$

The following nondimensional parameters were used: $\Omega = \omega[I/(MgR)]^{1/2}, \tau = t[MgR/I]^{1/2}, I = (4/3)MR^2, D = MR^2/I = 3/4$. The signum function $S$ is such that $S = +1$ for $\theta + \alpha_0 > 0$ and $S = -1$ for $\theta + \alpha_0 < 0$. In practice the equation of motion is integrated starting from the initial conditions $\theta(0) = 0, \theta' = \alpha_0'(0) = -\psi\Omega$, where a prime denotes differentiation with respect to scaled time $\tau$.

It is convenient to label the relative angle between the block and the base as $\Theta(\tau) = \theta(\tau) + \alpha_0(\tau)$. It is important from a numerical stability standpoint to make sure that the condition $\Theta = 0$ is accurately captured (otherwise a cumulative numerical error can occur). This is the switching point and represents the sudden transition between rocking about one corner to rocking about the other. It is also at this point that a discrete loss of energy can be incorporated via a coefficient of restitution.

There are a number of subtle modeling issues concerning this type of system. Much of the previous work has focused on a block rocking on a base that oscillates horizontally [10]. There is a discrete loss of energy associated with conservation of angular momentum as well as issues concerning the smooth initiation of rocking (and its dependence on an initial forcing phase). Further details can be found in Virgin et al. [7].

## 3 SOME SAMPLE NUMERICAL RESULTS

A typical output from the numerical simulation is shown in Fig. 2. In this case the forcing parameters are $\psi = 0.025, \Omega = 6.0$. Here, the seesawing base excitation (a sine wave) is switched on and persists for 15 cycles (part (c)). Part (a) shows the

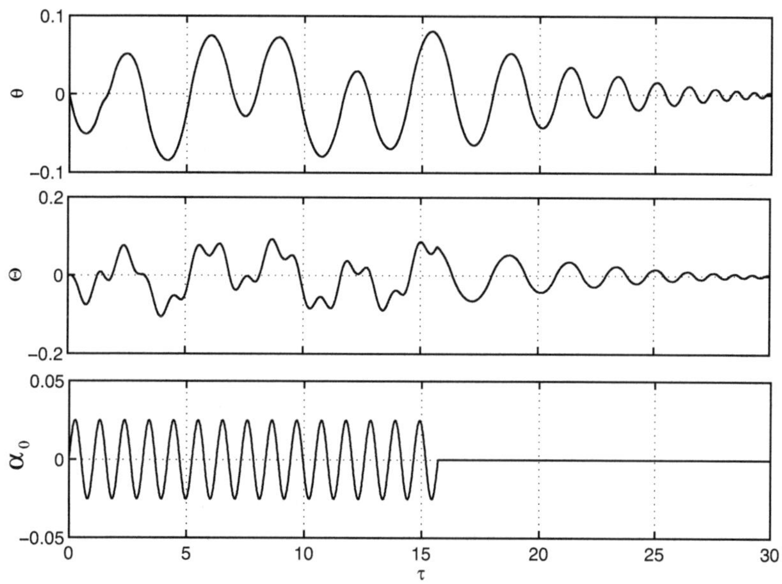


**Figure 2:** *A typical (non-overturning) time series. (a) absolute angle, (b) relative angle, (c) forcing function.*

response angle of rocking (in Radians) as a function of time. The *relative* angle is shown in part (b). The block undergoes a free rocking decay after the forcing is turned off, and the block comes to rest. In this specific case the block does not overturn.

However, if the forcing parameters are changed, for example, to $\psi = 0.06, \Omega = 1.84$, we can get an overturning event as shown in Fig. 3, and this occurs after about eight cycles of forcing.

The rocking response can also be usefully viewed in terms of a phase projection. Plotting angle versus angular velocity reveals the periodic (or otherwise) nature of the response. For example, Fig. 4 is the same data as in Fig. 2 but plotted as a phase projection in terms of the absolute angle in part (a) and relative angle in part (b).

Similarly the overturning response can also be plotted as a phase projection, see Fig. 5.

## 4 THE EXPERIMENTAL SET-UP

In order to verify the type of rocking behavior observed in the numerical simulations a simple mechanical system was devised. This is shown in Fig. 6. This experimental set-up is in the early stages of development and hence only a *qualitative* preliminary result is presented. An aluminum block measuring 6 by 2 inches was placed on a rigid base (note that this is slightly different from the 4 to 1 aspect ratio used in the simulations). One end of the base was attached (via a low-friction pivot) to a rotating link that was itself attached to the shaft of a motor. Although the block is modeled as a two-dimensional object the third dimension was used (together with guides) to promote planar motion. The base of the block was slightly concave to ensure pure rocking about either corner, and sandpaper was attached to the base to minimize sliding of the block.

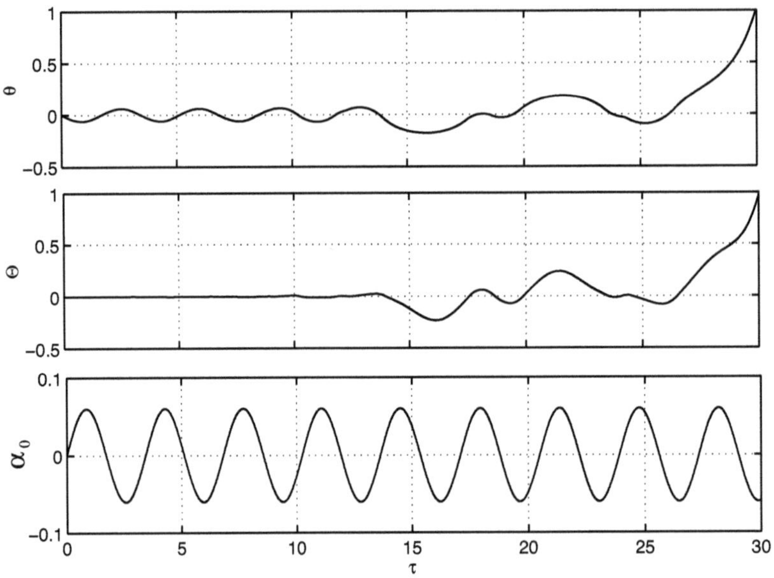


**Figure 3:** *A typical (overturning) time series. (a) absolute angle, (b) relative angle, (c) forcing function.*

A target was placed at the geometric center of the block as well as at a specific point on the seesawing base. These were the locations where the video camera and associated image processing software took data and provided time series of absolute (a fixed reference frame) and relative (the block with respect to the base) angles. The speed of the motor was used to change the frequency of the angular motion of the base. The lever arm could be changed to alter the forcing magnitude (there is a slight deviation from a pure harmonic) and the speed of the motor was used to obtain different forcing frequencies.

Fig. 7(a) shows the camera mounted on the isolation table together with the block and a target located at its centroid. The view in part (b) is an image of what the camera acquires. The target is sufficiently contrasted with the background that it (or rather the geometric center of its shape) can be tracked as a function of time. The camera was able to take about 200 frames per second.

## 5 PRELIMINARY EXPERIMENTAL RESULTS

Fig. 8 shows a typical output. Here, the target is tracked and plotted as a phase trajectory $(x - y)$ in terms of pixel location. The initial transient is clearly observable before it settles onto the steady state. The spread in the data is most likely due to a very small amount of sliding during the motion. Recall that the theoretical model assumes that the block does not slide, bounce or move out of plane. It is also interesting to note a slight asymmetry in the response. This is most likely the result of the motor, and hence the tilting platform, spending half the time in the same direction as gravity, and half against it.

In Fig. 9 is shown the same data but in terms of x-position and y-position time series. The axes are still in terms of pixels, and note that the x-position scale is somewhat larger than the y-position scale. The sudden reversal in direction as the block switches from rocking about one corner to the other is most clearly seen in the vertical position. The time scale in these figures is milliseconds. The asymmetry in alternating amplitudes can also be seen. These time series are in steady-state, i.e., an initial transient phase is not plotted. In order to compare numerical simulation with experimental data, the x-y location of the centroid of the block can then be converted to the angle. This will form the basis of further study.

## 6 CONCLUSIONS

A rigid block is an example of a single-degree of freedom non-smooth system. The nonlinearity may be considered severe and occurs discretely when the blocks switching from rocking about one corner to rocking about the other. For large angle of

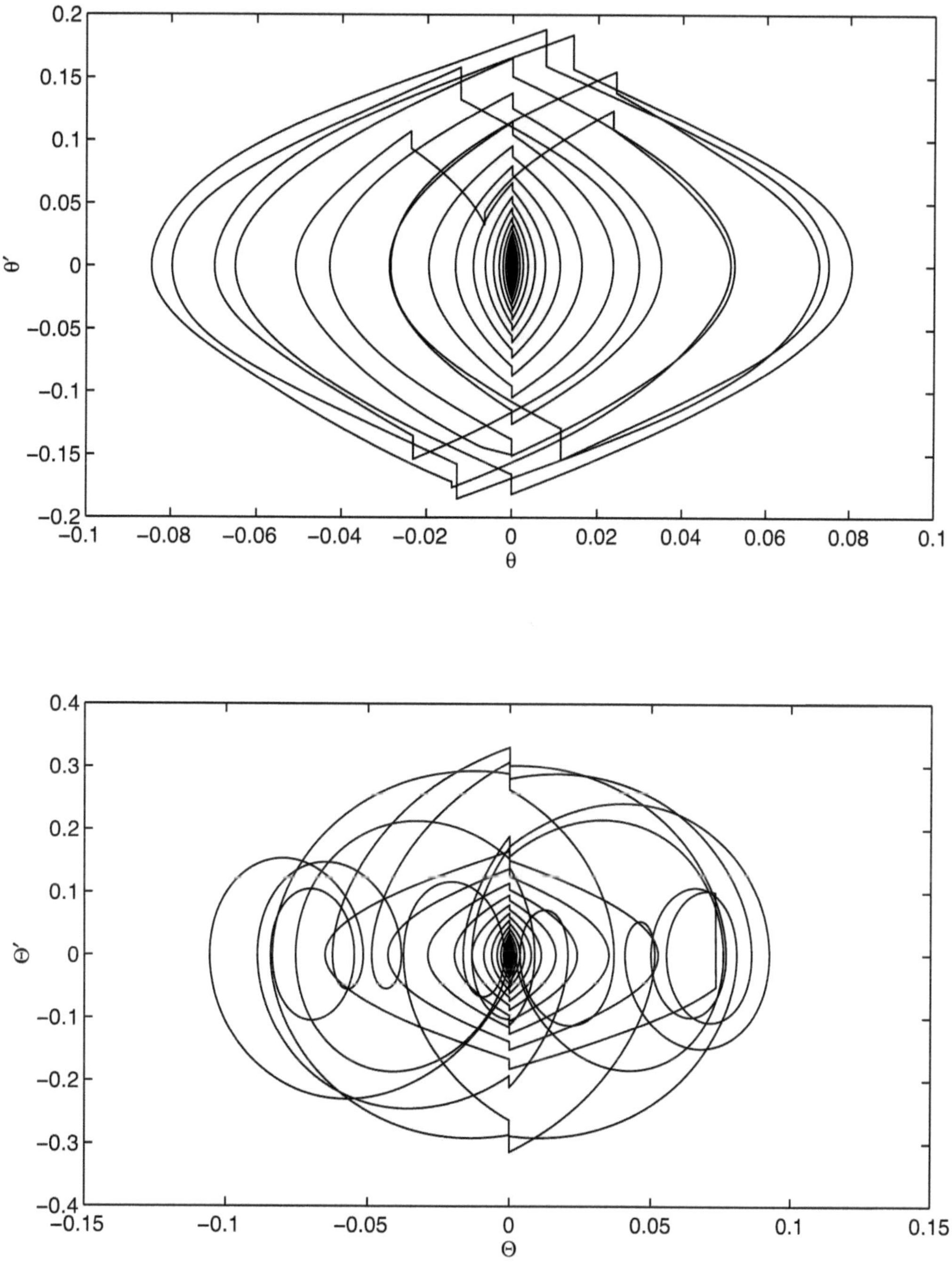


**Figure 4:** *A typical (non-overturning) phase projection. (a) absolute angle, (b) absolute angular velocity, (c) relative angle, (d) relative angular velocity, (e) forcing function.*

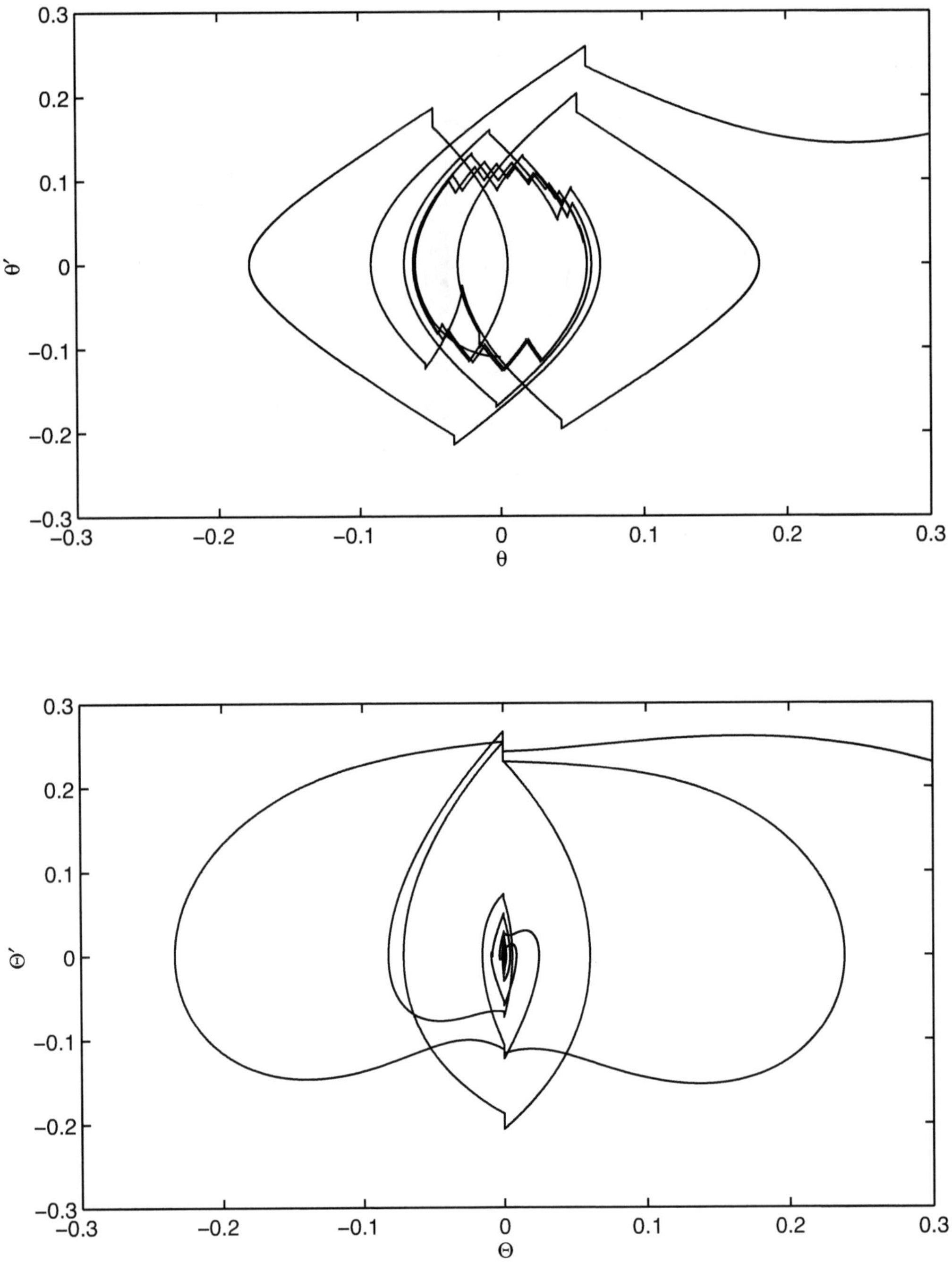


**Figure 5:** *A typical (overturning) phase projection. (a) absolute angle, (b) absolute angular velocity, (c) relative angle, (d) relative angular velocity, (e) forcing function.*

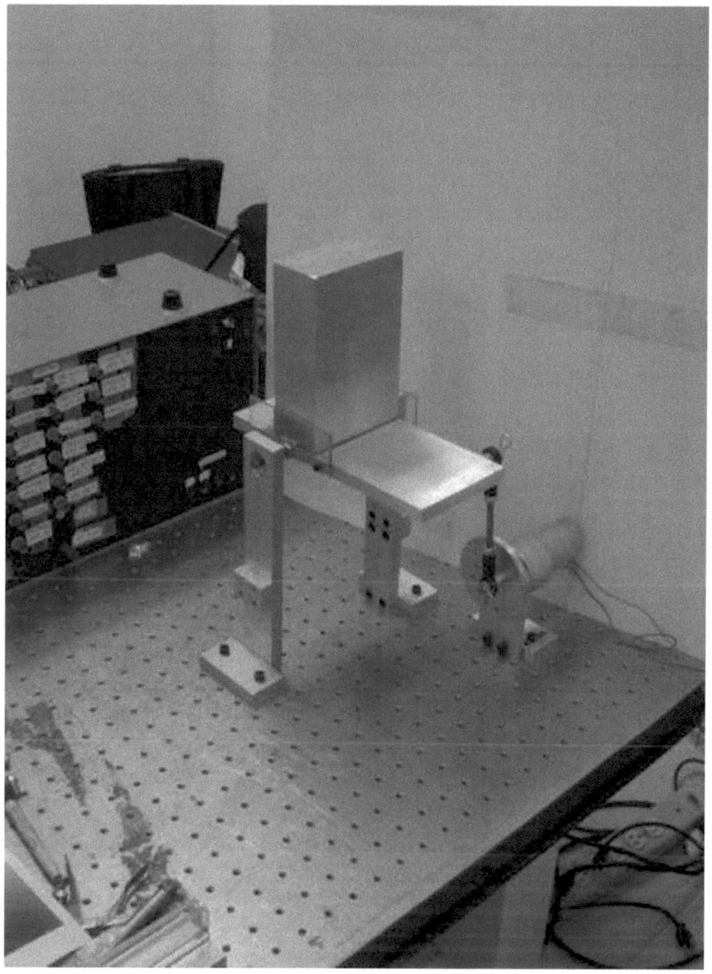

**Figure 6:** *The experimental system.*

**Figure 7:** *(a) the camera-based data acquisition system, (b) what the camera sees.*

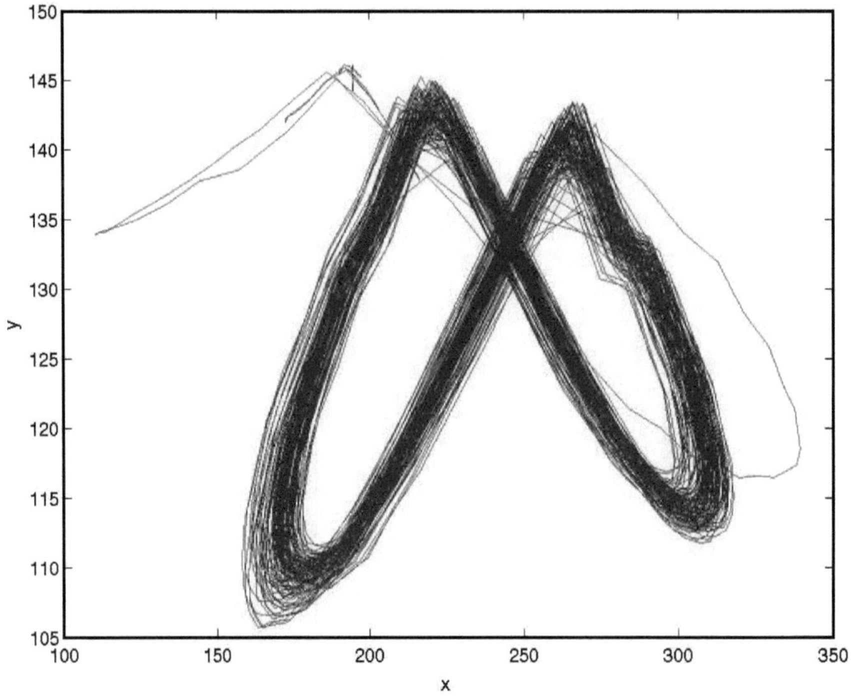


**Figure 8:** *A typical phase projection from the experiment.*

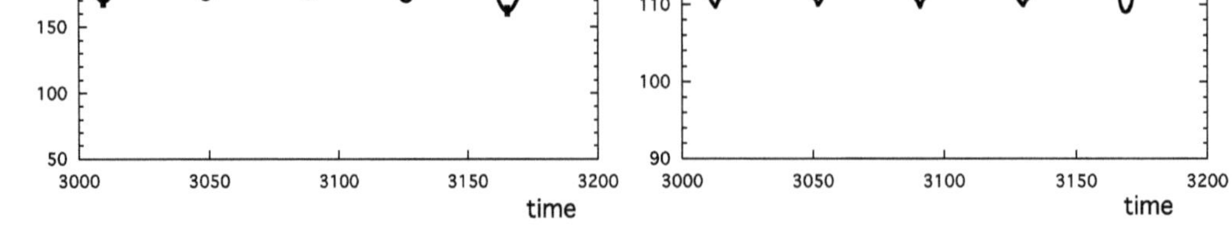


**Figure 9:** *Time series from the experiment.*

rocking (and depending on the slenderness of the block) there may be some geometric nonlinearity which can ultimately lead to overturning and the complete loss of stability. This non-smooth system is typical of a wide class of oscillator including mechanical systems with impact and Coulomb damping. The primary focus of this study was to obtain preliminary data from an experiment and confirm some of the interesting and difficult to predict transient behavior of a rocking block.

*Acknowledgement:* the assistance of Mukarram Ahmad,a Duke undergraduate, with the experimental data acquisition is gratefully acknowledged.

## REFERENCES

[1] **Virgin, L.**, Introduction to Experimental Nonlinear Dynamics, Cambridge University Press, 2000.

[2] **Housner, G.**, *The behavior of inverted pendulum structures during earthquakes*, Bulletin of the Seismological Society of America, Vol. 53, pp. 403–417, 1963.

[3] **Oliveira, C.** and **Bolt, B.**, *Rotational components of surface strong ground motion*, Earthquake Engineering and Structural Dynamics, Vol. 18, pp. 517–526, 1989.

[4] **Zembaty, Z., Castellani, A.** and **Boffi, G.**, *Spectral analysis of the rotational component of seismic ground motion*, Probabilistic Engineering Mechanics, Vol. 8, pp. 5–14, 1993.

[5] **di Bernardo, M., Budd, C., Champneys, A., Kowalczyk, P., Nordmark, A., Tost, G.** and **Piiroinen, P.**, *Bifurcations in Nonsmooth Dynamical Systems*, SIAM REVIEW, Vol. 50, No. 4, pp. 629–701, DEC 2008.

[6] **Shaw, S.**, *The Dynamics of a Harmonically Excited System Having Rigid Amplitude Constraints, Parts 1 and 2*, Journal of Applied Mechanics, Vol. 52, pp. 453–464, 1985.

[7] **Virgin, L., Fielder, W.** and **Plaut, R.**, *Transient motion and overturning of a rocking block on a seesawing foundation*, Journal of Sound and Vibration, Vol. 191, pp. 177–187, 1996.

[8] **Plaut, R., Fielder, W.** and **Virgin, L.**, *Fractal behavior of an asymmetric block overturning due to harmonic motion of a tilted base foundation*, Chaos, Solitons and Fractals, Vol. 7, pp. 177–196, 1996.

[9] **Fielder, W., Virgin, L.** and **Plaut, R.**, *Experiments and simulation of overturning of an asymmetric rocking block on an oscillating foundation*, European Journal of Mechanics A (Solids), Vol. 16, pp. 905–923, 1997.

[10] **Hogan, S.**, *On the dynamics of rigid block motion under harmonic forcing*, Proceedings of the Royal Society of London, Vol. A425, pp. 441–476, 1989.

Proceedings of the IMAC-XXVIII
February 1–4, 2010, Jacksonville, Florida USA


# Enhanced Order Reduction of Forced Nonlinear Systems Using New Ritz Vectors

Mohammad A. AL-Shudeifat, Eric A. Butcher and Thomas D. Burton
Department of Mechanical and Aerospace Engineering
New Mexico State University, Las Cruces, NM, 88003, USA
Email: eab@nmsu.edu

ABSTRACT

Enhanced modal-based order reduction of forced structural dynamic systems with isolated nonlinearity has been performed using the iterated LELSM (Local equivalent linear stiffness method) modes and new type of Ritz vectors. The iterated LELSM modes have been found via iteration of the modes of the mass normalized local equivalent linear stiffness matrix of the nonlinear systems. The optimal basis vector of principal orthogonal modes (POMs) is found for such system via simulation and used for POD-based order reduction for comparison. Two new Ritz vectors are defined as a static load vectors where one of them gives a static displacement to the mass connected to the periodic forcing load and the other gives a static displacement to the mass connected to the nonlinear element. It is found that the use of these vectors, which are augmented to the iterated LELSM modes in the order reduction modal matrix, reduces the number of modes used in order reduction and considerably enhances the accuracy of order reduction. The combination of the new Ritz vectors with the iterated LELSM modes in the order reduction matrix yields more accurate reduced models than POD-based order reduction of forced and nonlinear systems. Hence, the LELSM modal-based order reduction is essentially enhanced over POD-based and linear-based order reductions by using these new Ritz vectors. In addition, the main advantage of using the iterated LELSM modes for order reduction is that, unlike POMs, they do not require *a priori* simulation and thus they can be combined with new Ritz vectors and applied directly to the system.

## 1. Introduction

Various methods have been employed in approximating the response of the linear and nonlinear dynamic systems of $n$-dimensional space by $m$-dimensional subspace of coordinates. The classical method for order reduction of linear systems is due to Guyan [1]. Extensions of Guyan reduction have been proposed that include inertial as well as stiffness effects in the order reduction transformation [2]. These linear-based Guyan-like order reduction techniques have also been applied to nonlinear dynamical systems with weak static and damping nonlinearities [3-6].

Another method for order reduction is the invariant-manifold approach that has been used in reducing the dimension of the nonlinear dynamical systems to a reduced subspace of coordinates. It utilizes the nonlinear normal modes (NNMs) that describe the nonlinear motion of the system on a two-dimensional invariant-manifold in the system phase space [7-10]. As a result, the NNM-based reduced models are obtained in a subspace of the master coordinates [11-12]. However, the NNM-based reduced models become less accurate than the linear-based reduced models at the internal resonance condition. The results of the frequency-amplitude-dependence in [4] showed that the NNM-based reduced models were less accurate than the linear-based reduced models of the nonlinear systems with smooth nonlinearities for a range of parameters in the vicinity of an internal resonance condition. The invariant-manifold approach was also employed in time-periodic systems [13] in which the use is made of the Liapunov-Floquet transformation [14].

The principal orthogonal decomposition (POD) is also a well-known technique that is used in modal analysis of nonlinear structural systems [15-22]. POD-based order reduction requires *a priori* simulation of the dynamic system to form a subset of the dominant proper orthogonal modes (POMs) of the highest singular values. These optimal bases form the columns of the POD-based order reduction transformation matrix. The technique was applied for order reduction of the structural system with frictional excitation [19] and nonlinear dynamic systems with smooth and non-smooth nonlinearities [20-22]. The POD-based reduced model was considerably enhanced via augmentation of the optimal bases of the POMs with a new type of Ritz vectors called Kumar-Burton or K-B vectors for dynamic systems with coupling nonlinearities [22]. The Ritz vector approach has also been applied in [23] of which a single Ritz vector is employed for each isolated nonlinearity along with a subset of linear basis functions to obtain a more accurate reduced model than that obtained only with the linear functions. However, the use of Ritz vectors in enhancing the accuracy of order reduction of dynamic systems is not a new topic. The approach was previously applied in order reduction of linear forced dynamic systems [24-25]. The optimal Ritz vectors were augmented to a truncated subset of eigenvectors of the modal matrix of the linear system to form the order reduction transformation matrix. The load dependent Ritz vectors were found by assigning a spatial distribution of a dynamic load to each substructure of the system [26-27].

Most recently, a new technique for order reduction of dynamical systems with static piecewise linear nonlinearities was proposed based on previously developed methods for approximation of the NNM frequencies and mode shapes via either an amplitude-independent piecewise modal method (PMM) or an amplitude-dependent local equivalent linear stiffness method (LELSM) [6,28,29]. The technique was further improved over the previously developed methods for order reduction of large degree-of-freedom unforced nonlinear systems with grounded nonlinearities via iteration of the LELSM modes. The iteration of the LELSM modes was found to be a good approximation to the POMs of the systems. These iterated LELSM modes were found to be as efficient as POMs in the order reduction of such systems [29].

The LELSM modal-based order reduction via iterated LELSM modes is enhanced in this study by using a new type of Ritz vector: the Sh-B vector. This approach is applied to forced dynamic systems with static grounded and coupling nonlinearities. The new Sh-B vectors are augmented to the iterated LELSM order reduction matrix for such systems. These vectors considerably enhance the order reduction of these systems over the POD-based and linear-based order reductions.

## 2. Principal Orthogonal and Iterated LELSM modes

### 2.1 Proper Orthogonal Decomposition

Consider the forced and damped $n-dof$ dynamical system with static nonlinearity as shown in Figure 1.

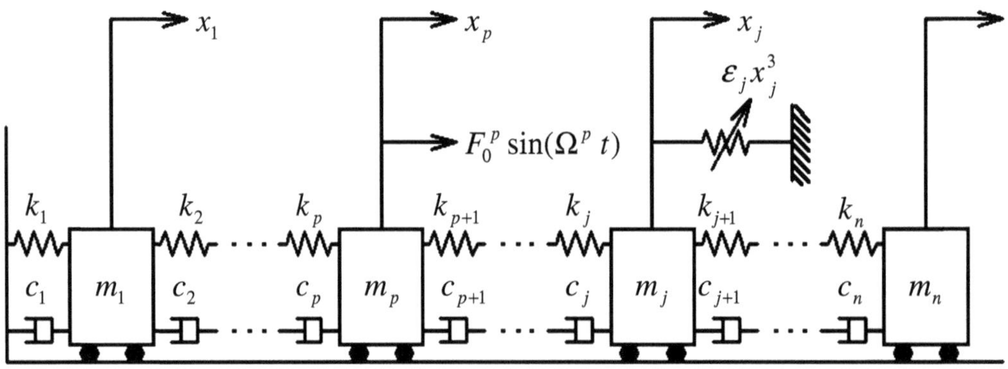


**Figure 1** $n-dof$ mass-spring system with forcing and one cubic spring.

The equations of motion of the system can be written as

$$M\,\ddot{x} + C\,\dot{x} + K\,x + F_{nl}(x) = F(t) \tag{1}$$

where $x \in \mathbf{R}^n$ is the vector of displacements, $M$ is the $n \times n$ mass matrix, $K$ is the $n \times n$ stiffness matrix, $C$ is the $n \times n$ damping matrix, $F_{nl}(x) = [0\ \cdots\ 0\ \ f_{nl}^{j}(x_j)\ \ 0\ \cdots\ 0]^T$ is the nonlinear $n \times 1$ force vector for $1 \le j \le n$ which consists of grounded static nonlinearities and $F(t) = [0\ \cdots\ 0\ \ F_0^p \sin(\Omega^p t)\ \ 0\ \cdots\ 0]^T$ is the $n \times 1$ periodic excitation force vector for $1 \le p \le n$. The method of proper orthogonal decomposition (POD) requires *a priori* simulation of the dynamic system in Equation (1) in space and time. The solution at *m* time steps is written in matrix form as [20]

$$X = [x_1\ \ x_2 \ldots x_n] = \begin{bmatrix} x_{11} & \cdots & x_{1n} \\ \vdots & \ddots & \vdots \\ x_{m1} & \cdots & x_{mn} \end{bmatrix} \tag{2}$$

The singular value decomposition is used for determining the POMs and the singular values $\sigma_i^2$ of the system. For the $m \times n$ matrix X in Equation (2) the SVD can be written as [20]

$$X = USV^T \tag{3}$$

where the POMs are the column vectors of the matrix V and the diagonal elements of matrix $S$ are called the singular values of the system. The subset of POMs that correspond to the higher singular values are selected to form the order reduction modal matrix. Selecting the POMs of higher singular values means selecting the POMs of the greatest amount of energy in the signal. The ratio of energy in the first $m$ POMs of highest singular values to the total energy of the $n$ POMs is given by [20]

$$R_{POD} = \frac{\sum_{i=1}^{m} \sigma_i^2}{\sum_{i=1}^{n} \sigma_i^2} \tag{4}$$

where $\sigma_i^2$ is the singular value corresponding to the mode $i$ and $\sigma_1^2 > \sigma_2^2 > \cdots > \sigma_n^2$.

### 2.2 Local Equivalent Linear Stiffness Method

The local equivalent linear stiffness method (LELSM) has been utilized previously in [6,28,29]. If $F_{nl}(x)$ in Equation (1) acts on some $jth-dof$ then setting $F(t)$, $C$ and all other degrees-of-freedom to zero yields

$$m_j\,\ddot{x}_j + k_j\,x_j + f_{nl}^{j}(x_j) = 0 \tag{5}$$

The nonlinear frequency of this single $dof$ system is $\Omega_j$. The stiffness matrix $K$ in Equation (1) is updated to $K_{eq}$ by updating the diagonal element $k_j$ as

$$k_j = (k_{eq})_j = m_j\,\Omega_j^2 \tag{6}$$

By setting all masses and stiffnesses to unity for the system in Figure 1, in an appropriate units, Equation (5) becomes

$$\ddot{x}_j + 2x_j + f_{nl}^{j}(x_j) = 0 \tag{7}$$

For $f_{nl}^{j}(x_j) = \varepsilon_j x_j^3$, Equation (7) can be rewritten as

$$\ddot{x}_j + \omega_j^2 \left( x_j + \frac{\beta_j \varepsilon_j}{\omega_j^2} x_j^3 \right) = 0 \tag{8}$$

where $\omega_j^2 = 2$ and $\beta_j = 1$. An approximation for the nonlinear frequency in Equations (8) is given by [30]

$$\left(k_{eq}\right)_j = \Omega_j^2 = \frac{\omega_j^2}{144}(80 + 62(\frac{\beta_j \varepsilon_j}{\omega_j^2})X_j^2 + \sqrt{4096 + 5888(\frac{\beta_j \varepsilon_j}{\omega_j^2})X_j^2 + 1684(\frac{\beta_j \varepsilon_j}{\omega_j^2})^2 X_j^4} \tag{10}$$

where $j$ is the degree of freedom location and $X_j$ is the initial amplitude at that $dof$.
As a result, the updated stiffness matrix is given by

$$K_{eq} = \begin{bmatrix} 2 & -1 & & & & & & \\ -1 & 2 & -1 & & & & 0 & \\ & \ddots & \ddots & \ddots & & & & \\ & & -1 & \left(k_{eq}\right)_j & -1 & & & \\ & & & \ddots & \ddots & \ddots & & \\ & 0 & & & -1 & 2 & -1 \\ & & & & & -1 & 1 \end{bmatrix} \tag{11}$$

The local equivalent stiffness matrix is then normalized as $\left[\tilde{K}_{eq}\right] = [M]^{-1/2}\left[K_{eq}\right][M]^{-1/2}$. Hence, the eigen-value problem $\left|\tilde{K}_{eq} - \Omega_i^2 I\right| \mathrm{x} = 0$ yields the LELSM modes and frequencies. It is clear that, unlike POD, the LELSM does not require *a priori* simulation of the dynamical system in Equation (1). In the following, the LELSM methodology is modified by iterating the LELSM modes. If the $n-dof$ system has the $jth-dof$ attached to a nonlinear spring then the iterated LELSM modes are found via the following iteration technique:

1. Use arbitrary initial amplitude for $x_j^{(1)}(0)$ or let $x_j^{(1)}(0) = \phi_{j1}$ where $\phi_{j1}$ is the $jth$ element of the first linear mode of the system.
2. At $k = 1$ and $x_j^{(1)}(0)$ find $\left(k_{eq}\right)_j$ to form $\left(K_{eq}\right)_{n\times n}$ then solve for the first LELSM mode shape $\tilde{\phi}_1^{(1)}$ to get $\tilde{\phi}_{j1}^{(1)}$.
3. At $k = 2$, set $x_j^{(2)}(0) = (1/2)\left(\tilde{\phi}_{j1}^{(1)} + x_j^{(1)}(0)\right)$ and solve for $\tilde{\phi}_1^{(2)}$ to get $\tilde{\phi}_{j1}^{(2)}$.
4. For each new value of $x_j^{(k)}(0)$, find $\left(k_{eq}\right)_j$ to form $\left(K_{eq}\right)_{n\times n}$ and solve for a new $\tilde{\phi}_1^{(k)}$ to get new value of $\tilde{\phi}_{j1}^{(k)}$.
5. Continue iteration up to $k = m$, then set $x_j^{(m)}(0) = (1/2)\left(\phi_{j1}^{(m-1)} + x_j^{(m-1)}(0)\right)$ and solve for $\tilde{\phi}_1^{(m)}$. If $\tilde{\phi}_1^{(m)} - \tilde{\phi}_1^{(m-1)} \le e$, then stop. Otherwise continue iteration until the LELSM mode converges to a unique shape. Hence, the iterated LELSM modes are obtained.

It is found that the POMs obtained by using the first linear mode as the initial condition for the unforced nonlinear system are well-approximated by the iterated LELSM modes [29].

## 3. New Ritz Vectors for Order Reduction

The iterated LELSM modes have been found as efficient as POMs in reducing the order of unforced and damped nonlinear dynamical system with static nonlinearities [29]. The use of the iterated LELSM modes in reducing the order of such systems with a periodic forcing excitation, such as the system in Figure 1, is greatly aided through the simultaneous use of new Ritz vectors that enhance the order reduction. For this purpose, two new Ritz vectors are found and combined with the iterated LELSM modes in the order reduction modal matrix. The first static load Ritz vector $R_1$ is required to give the forced mass a unity static displacement, while all other masses have zero displacements, which yields the displacement vector $r_1$. The second load Ritz vector $R_2$ is required also to give the mass attached to the cubic spring a unity static displacement while all other masses have zero displacements, which yield the displacement vector $r_2$ . Hence, using the equivalent linear stiffness matrix which takes the nonlinearities into account, we have

$$R_1 = K_{eq}\, r_1 \tag{12a}$$

$$R_2 = K_{eq}\, r_2 \tag{12b}$$

These new Ritz vectors are the Sh-B vectors.

## 4. Modal-Based Order Reduction Methodologies

It is desired to reduce the order of the forced, damped and nonlinear $n-dof$ mass-spring system that shown previously in Figure 1 to an equivalent $m-dof$ system where $m << n$ . Hence, the modal based order reduction requires forming a $n \times m$ matrix $\tilde{\Phi}$ of the first $m$ columns of the $n \times n$ modal matrix of the POD modes, LELSM modes or the linear modes of the system as

$$\tilde{\Phi}_{n\times m} = \left[\tilde{\phi}_1 \;\; \tilde{\phi}_2 \cdots \tilde{\phi}_m\right] \tag{13}$$

The order reduction is performed by applying the transformation $x_{n\times 1} = \tilde{\Phi}_{n\times m}\, z_{m\times 1}$ to Equation (1) and premultiplying by $\tilde{\Phi}^T$ which yields

$$\overline{M}\,\ddot{z} + \overline{C}\,\dot{z} + \overline{K}\,z + f_{nl}(z) = f(t) \tag{14}$$

where $\tilde{\Phi}^T M\, \tilde{\Phi} = \overline{M}$ , $\tilde{\Phi}^T C\, \tilde{\Phi} = \overline{C}$ , $\tilde{\Phi}^T K\, \tilde{\Phi} = \overline{K}$ , $\hat{F}(z) = F_{nl}\left(\tilde{\Phi}\, z\right)$ , $f_{nl}(z) = \tilde{\Phi}^T \hat{F}(z)$ , and $f(t) = \tilde{\Phi}^T F(t)$. To enhance the modal-based order reduction, the new Sh-B displacement vectors $r_1$ and $r_2$ are augmented to the order reduction modal matrix $\tilde{\Phi}$ as

$$\tilde{\Phi}_{n\times m} = \left[\tilde{\phi}_1 \;\; \tilde{\phi}_2 \;\cdots\; \tilde{\phi}_{m-2} \;\; r_1 \;\; r_2\right] \tag{15}$$

This new order reduction modal matrix is used in Equation (14) to obtain an enhanced order reduction via Sh-B vectors.

## 5. Order Reduction of Forced, Damped and Nonlinear $20-dof$ Mass-Spring System

For the forced, damped and nonlinear system in Figure 1 , all masses and stiffnesses are set to unity at an appropriate units, $n = 20$ , $F_0^p = 3$ , $\Omega^p$ is the average of the first three natural frequencies of the linear system and the damping matrix is proportional to the stiffness matrix as $C = \gamma K$ where $\gamma = 0.04$ . Two possible cases are considered as shown in Table 1 of the forcing and the static nonlinearity locations to show the capability of these new Sh-B vectors in enhancing the order reduction of the system. The initial condition for all cases is the first linear mode $x(0) = \phi_1$ for the displacements and $\dot{x}(0) = 0.1\,\phi_1$ for the velocities.

Table I Possible cases for forcing and nonlinearity locations

| | | |
|---|---|---|
| Case 1 | $p = 4$ | $j = 16$ |
| Case 2 | $p = 4$ | $j = n = 20$ |

Case 1: In this case the periodic excitation load acts on $m_4$ $(p=4)$ and the cubic spring is attached to $m_{16}$ $(j=16)$ where $f_{nl}^{16}(x_{16})=\varepsilon_{16}x_{16}^3$ and $\varepsilon_{16}=3$. The first static load Sh-B vector $R_1$ is used to displace $m_4$, at which the forcing excitation acts, by a unit displacement and keeps all other masses with zero displacements. The second static load Sh-B vector $R_2$ is used to displace $m_{16}$ by a unit displacement and keeps all other masses with zero displacements. Figure 2 shows the necessary static Sh-B load vectors and their corresponding static displacement vectors of the system.

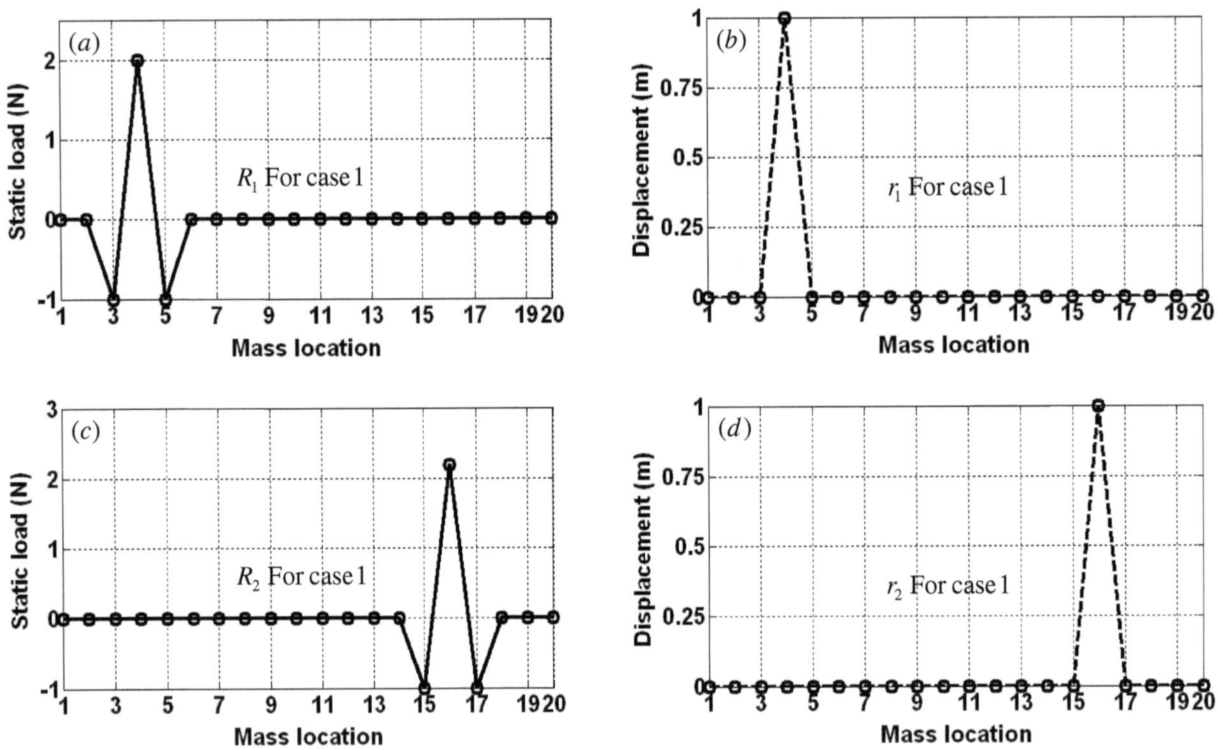


**Figure 2** Static load Sh-B vectors (new Ritz vectors) and their corresponding displacements of masses of the $20-dof$ system with forcing excitation at $m_4$ and cubic spring attached to $m_{16}$ for case 1.

The time histories of the mass $m_{16}$ in Figure 3 show the use of the iterated LELSM modes without Ritz vectors is not accurate in reducing the order of the system even high number of modes is retained. In figure 4 the augmentation of the Sh-B displacement vectors $r_1$ and $r_2$ in the order reduction modal matrix considerably enhanced the reduced order models while the use of 10 iterated LELSM modes combined with two Sh-B vectors gives a simulation of a reduced order model that is nearly equivalent to the full model simulation.

To explore more evidences that the new Sh-B vectors are enhancing the LELSM modal-based order reduction over the POD-based order reduction, the phase plane portraits have been plotted for $m_{16}$ of the system. Figure 5b shows that the LELSM modal-based reduced model that uses 10 iterated LELSM modes combined with the Sh-B vectors is much enhanced over the reduced model in Figure 5a that uses 12 iterated LELSM modes without Sh-B vectors. The use of the Sh-B vectors with POMs does not give an accurate reduced model as shown in Figure 6b when 10 POMs are retained and combined with two Sh-B vectors. The reduced model in Figure 5b that uses 10 iterated LELSM modes and two Sh-B vectors is little more accurate than the reduced model in Figure 6a that uses 12 POMs. Hence, the new Sh-B vectors enhance the LELSM modal-based reduced models over the POD-based reduced models. In addition, the use of these Sh-B vectors with the linear modes considerably enhanced the linear-based order reduction as shown in Figure 7.

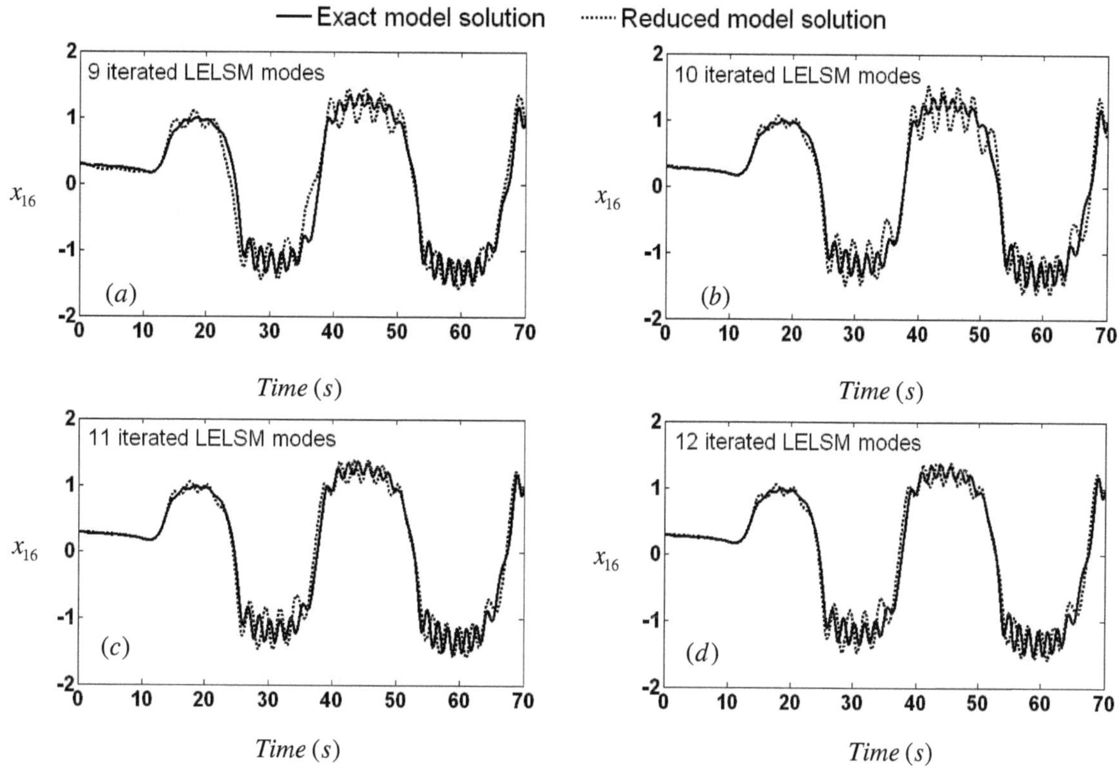


**Figure 3** Simulations of the forced 20-dof system in case 1 and its reduced models, (a) 9 iterated LELSM modes, (b) 10 iterated LELSM modes, (c) 11 iterated LELSM modes, (d) 12 iterated LELSM modes.

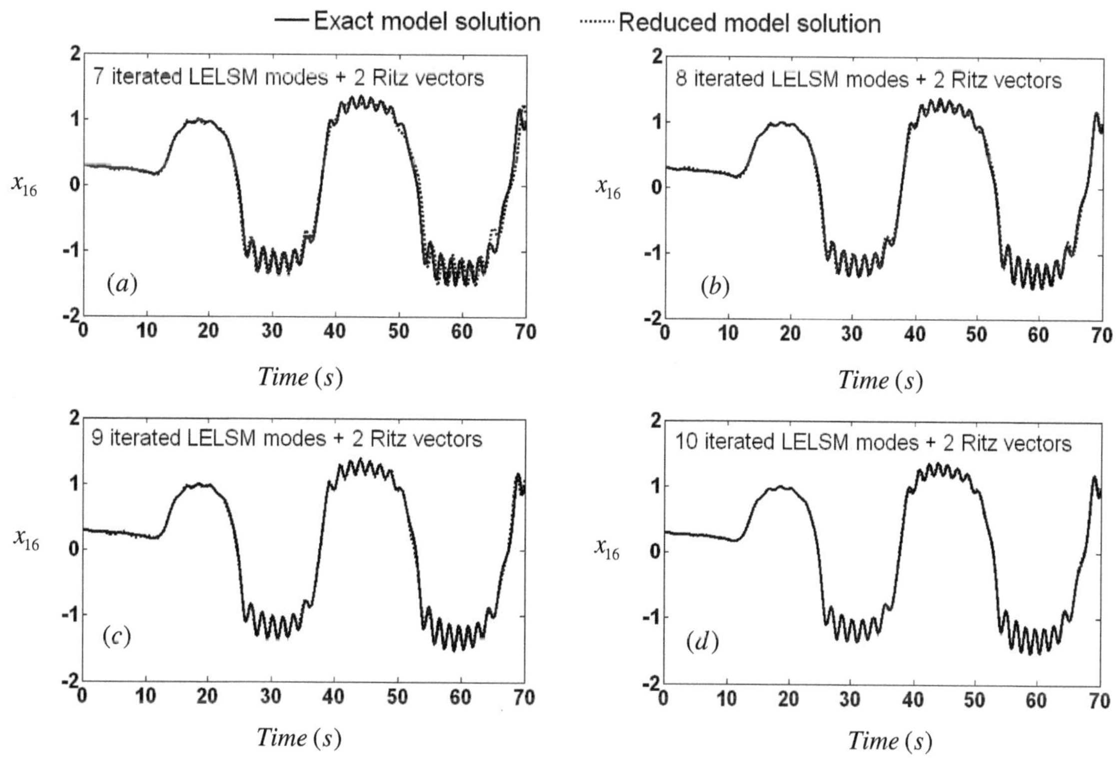


**Figure 4** Simulations of the forced 20-dof system in case 1 and its reduced models, (a) 7 iterated LELSM modes combined with 2 Sh-B vectors, (b) 8 iterated LELSM modes combined with two Sh-B vectors, (c) 9 iterated LELSM modes combined with 2 Sh-B vectors, (d) 10 iterated LELSM modes combined with two Sh-B vectors.

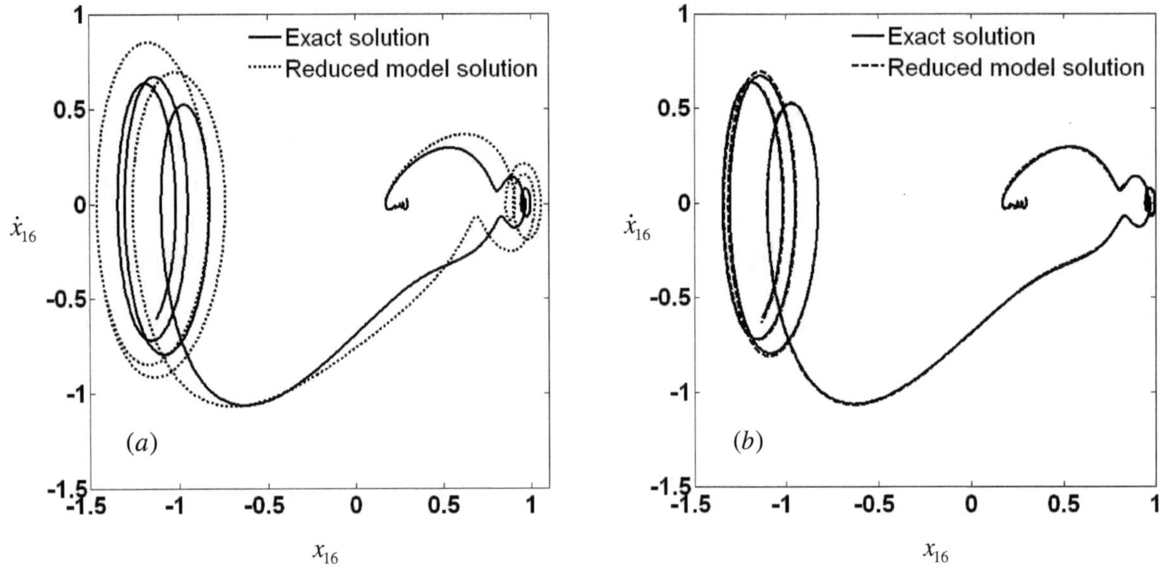


**Figure 5** Phase plane portraits that obtained via simulation of the system in case 1 up to $t = 35s$, (a) 12 iterated LELSM modes, (b) 10 iterated LELSM modes combined with two Sh-B vectors.

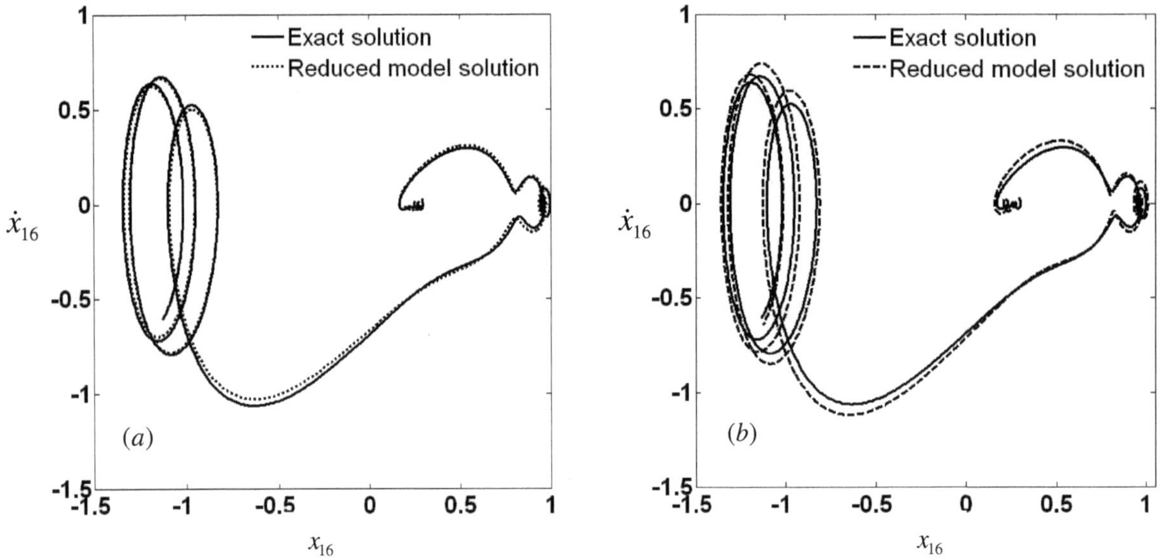


**Figure 6** Phase plane portraits that obtained via simulation of the system in case 1 up to $t = 35s$, (a) 12 POMs, (b) 10 POMs combined with two Sh-B vectors.

Moreover, the time histories of $x_4$ and $x_{16}$ are simulated up to $1000\,s$ as shown in Figure 8. It is clear that the LELSM modal-based reduced model that obtained by using 10 iterated LELSM modes and the two Sh-B vectors stays very accurate for long time period of simulation.

Reduced models via iterated LELSM modes and two Sh-B vectors are obtained for the same system for different values of the nonlinearity coefficients at fixed value of the periodic forcing amplitude as shown in Figure 9. In addition, Figure 10 shows the reduced models of the system at fixed nonlinearity coefficient for different values of the periodic forcing amplitudes. Both figures show that the LELSM modal-based order reduction via iterated LELSM modes and Sh-B vectors are accurate for different values of either forcing amplitudes or nonlinearity coefficients.

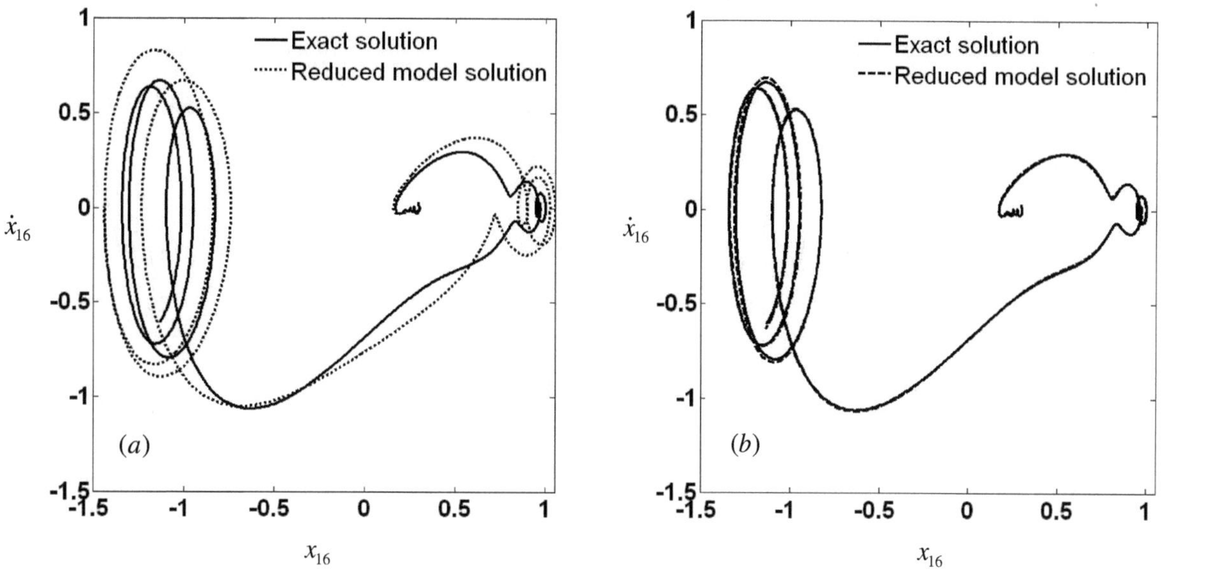


**Figure 7** Phase plane portraits that obtained via simulation of the system in case 1 up to $t = 35s$ , (a) 12 linear modes, (b) 10 linear modes combined with two Sh-B vectors.

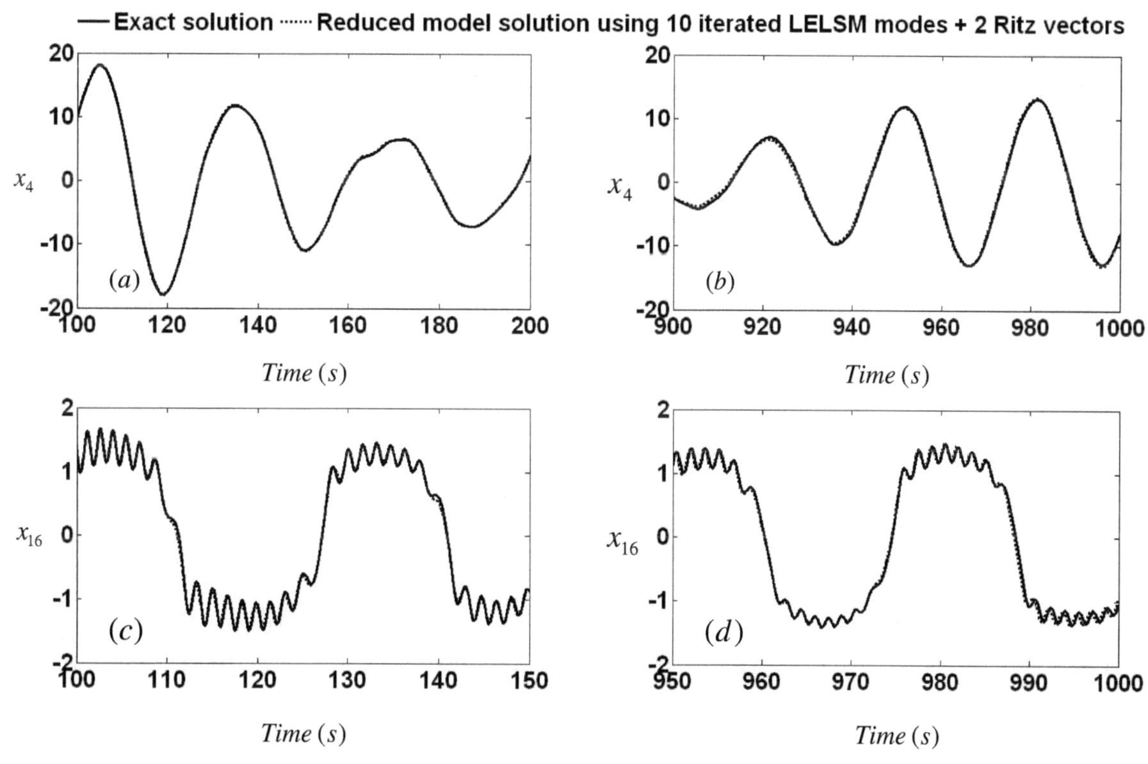


**Figure 8** Simulations of the forced 20-dof system in case 1 and its reduced models, (a) and (b) the 4[th] degree-of-freedom that attached to forcing, (c) and (d) the 16[th] degree-of freedom that attached to the cubic spring.

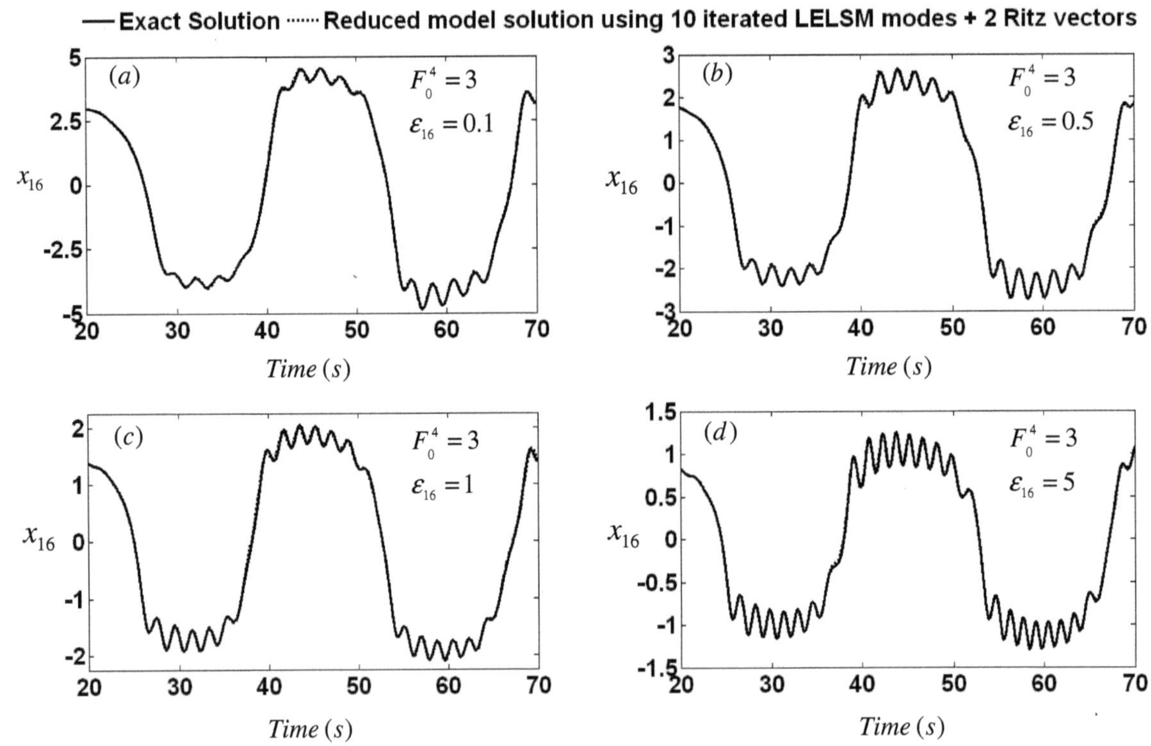


**Figure 9** Simulations of the 16[th] $dof$ of the full model and its reduced models for different values of the nonlinearity coefficient $\varepsilon$ for fixed value of periodic forcing amplitude $F_0^4$.

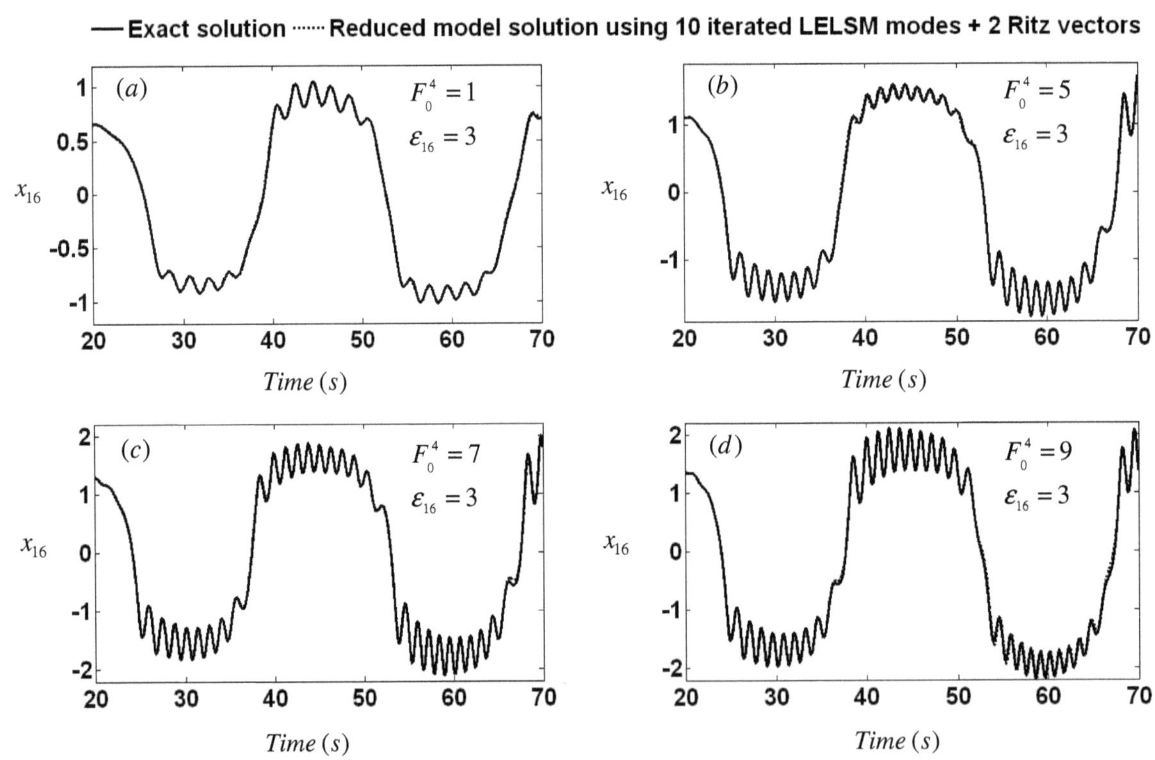


**Figure 10** Simulations of the 16[th] $dof$ of the full model and its reduced models for different values of the periodic forcing amplitudes $F_0^4$ for fixed nonlinearity coefficient $\varepsilon$.

**Case 2:** In this case we explore the capability of the new Sh-B vectors in enhancing the LELSM modal-based reduced model when the nonlinear spring is attached to the right end mass $m_{20}$ where $(j = n = 20)$, $f_{nl}^{20}(x_{20}) = \varepsilon_{20} x_{20}^3$ and $\varepsilon_{20} = 3$ and the periodic excitation load acts on $m_4$ $(p = 4)$. As before, the first static load Sh-B vector is used to displace $m_4$ by a unit displacement and keeps all other masses with zero displacements. The second static load Sh-B vector is used to displace $m_{20}$ by a unit displacement and keeps all other masses with zero displacements. The simulation results of the phase plane portraits for the full model and its reduced models are plotted for mass $m_4$ and mass $m_{16}$ in Figure 11. It is clear that the LESLM modal-based reduced model, that enhanced by the new Sh-B vectors, is more accurate than POD-based reduced model for $m_4$ phase plane portraits and it is nearly as accurate as the POD-based reduced model for $m_{20}$.

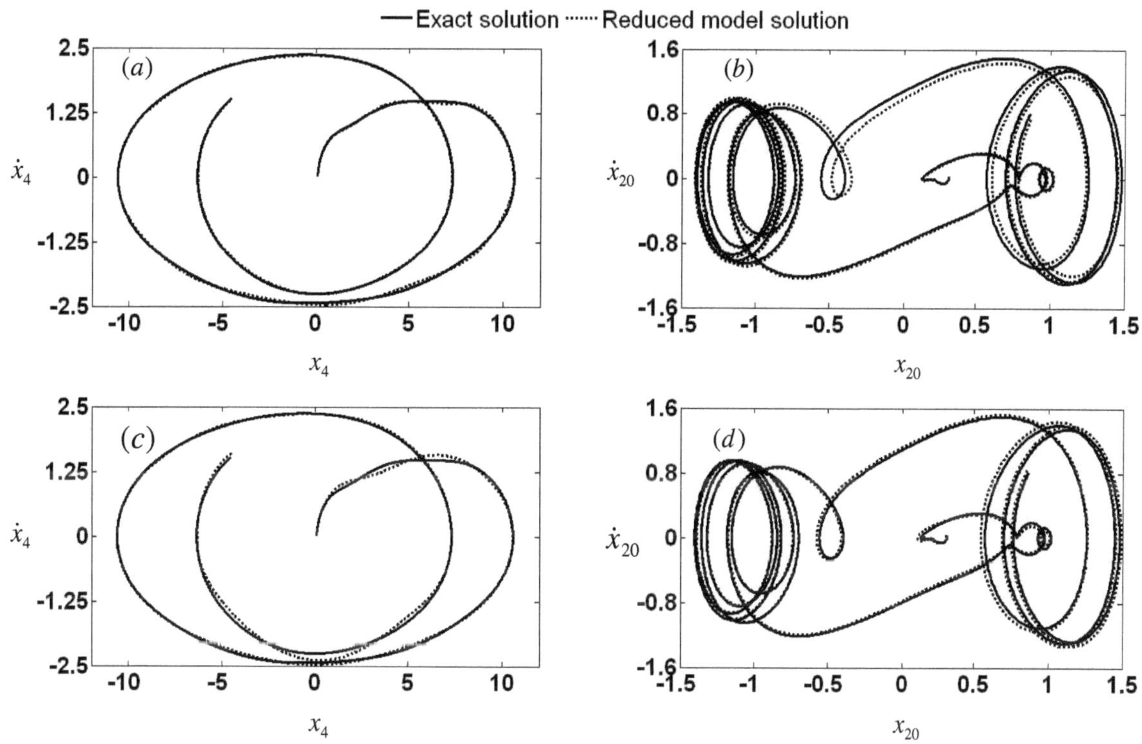


**Figure 11** Phase plane portraits that obtained via simulation of the system in case 2 up to $t = 120\,s$ , (a) and (b) 10 iterated LELSM modes combined with 2 Sh-B vectors, (c) and (d) 12 POMs.

## 6. Conclusions

An efficient order reduction technique for forced nonlinear dynamic systems with isolated nonlinearities is introduced in this paper. The technique is based on using a subset of the iterated LELSM modes augmented with new Ritz vectors (Sh-B vectors). Unlike linear modes, the nonlinearity effect appears in the iterated LELSM modes. These new modes preserve the nonlinear dynamics of the system. Hence, their use in order reduction for forced nonlinear systems was proven to be more efficient than POD-based and linear-based order reduction in this study.

The augmentation of the new Sh-B vectors with a subset of iterated LELSM modes was found more accurate in reducing the order of a forced system with grounded nonlinearities than the use of a subset of POMs of the same dimension. It was found that the optimal locations of the displacements of the new Sh-B vectors are the masses which are connected to the nonlinear spring and forcing. In addition, the augmentation of Sh-B vectors with a subset of POMs was found to be efficient in reducing the order of the forced dynamic system with coupling nonlinearity.

It is shown that Sh-B vectors can be used for different locations of nonlinear springs and forcing and yield accurate reduced models. The main advantage of using LELSM-based order reduction with Sh-B vectors is that, unlike POD-based order reduction, no *a priori* simulation of the full model is required and thus it can be applied directly to the model.

## References


[1] Guyan R. J., Reduction of stiffness and mass matrices, AIAA Journal, 2, 380, 1965.

[2] Burton T. D. and Young M.E., Model Reduction and Nonlinear Normal Modes in Structural Dynamics, Nonlinear and Stochastic Dynamics Symposium, AMD-Vol. 192, DE-Vol. 78, pp. 9 - 16, ASME Winter Ann. Mtg., Chicago, Nov. 6-11,1994.

[3] Friswell M. I., Penny J. E. T. and Garvey S. D., Using linear model reduction to investigate the dynamics of structures with local nonlinearities, Mechanical Systems and Signal Processing, 9(3), 317-328, 1995.

[4] Burton T. D. and Rhee W., On the reduction of nonlinear structural dynamics models, J. of Vibration and Control, 6, 531-556, 2000.

[5] Kim J. and Burton T. D., Reduction of structural dynamics models having nonlinear damping, J. Vibration and Control, 4,147-169, 2006.

[6] Butcher E. A. and Lu R., Order reduction of structural dynamic systems with static piecewise linear nonlinearities, Nonlinear Dynamics, 49, 375-399, 2007.

[7] Shaw S.W and Pierre C., Non-linear normal modes and invariant manifolds, J. Sound and Vibration, 150, 170-173, 1991.

[8] Jiang D., Pierre C. and Shaw S., Large amplitude nonlinear normal modes of piecewise linear systems, J. Sound and Vibration, 272, 869-891, 2004.

[9] Peschek E., Boivin N. and Pierre C., Nonlinear modal analysis of structural systems using multi-mode invariant manifolds, Nonlinear Dynamics, 25, 183-205, 2001.

[10] Burton T.D., Numerical Calculation of Nonlinear Normal Modes in Structural Systems, Nonlinear Dynamics, 49, 425 – 441, 2007.

[11] Shaw S. W., Pierre C. and Pesheck E., Modal analysis-based reduced-order models for nonlinear structures – An invariant manifold approach, Sound and Vibration Digest, 31, 3-16, 1999.

[12] Pesheck E., Pierre C. and Shaw S. W., Modal reduction of a nonlinear rotating beam through nonlinear modes, Journal of vibration and Acoustics, 124, 229-236, 2002.

[13] Sinha S. C., Redkar S. and Butcher E. A., Order reduction of nonlinear systems with time periodic coefficients using invariant manifolds, J. Sound & Vibration, 284, 985-1002, 2005.

[14] Sinha, S. C., Butcher E. A. and Dávid A., Construction of dynamically equivalent time invariant forms for time periodic systems, Nonlinear Dynamics, 16, 203-221, 1998.

[15] Feeny B. F. and Kappagantu R., On the physical interpretation of proper orthogonal modes in vibrations, Journal of Sound and Vibration, 211 (4), 607-616, 1998.

[16] Feeny B. F., On proper orthogonal co-ordinates as indicators of modal activity, Journal of Sound and Vibration, 255 (5), 805-817, 2002.

[17] Han S. and Feeny B. F., Enhanced proper orthogonal decomposition for the modal analysis of homogeneous structures, Journal of Vibration and Control, 8(1), 19-40, 2002.

[18] Lenaerts V., Kerschen G., Golinval J. C., Proper orthogonal decomposition for model updating of nonlinear mechanical systems, Mechanical Systems and Signal Processing, 15(1), 31-43, (2001).

[19] Kappagantu R. and Feeny B. F., An" optimal" modal reduction of a system with frictional excitation, Journal of Sound and Vibration, 224 (5), 863-877, 1999.

[20] Kerschen G., Golival J., Vakakis A. and Bergman L., The method of proper orthogonal decomposition for dynamical characterization and order reduction of mechanical systems: an overview, Nonlinear Dynamics, 41,147-169, 2005.

[21] Kumar N. and Burton T.D., Use of random excitation to develop POD based reduced order models for nonlinear structural dynamics, Proc. ASME IDETC, Paper DETC2007/VIB-35539, Las Vegas, NV, September 4-7 (2007).

[22] Kumar N. and Burton T.D., On Combined Use of POD modes and Ritz vectors for model reduction in nonlinear structural dynamics, Proc. ASME IDETC, Paper DETC2009-87416, San Diego, CA, September (2009).

[23] Segalman J. D., Model reduction of systems with localized nonlinearities, Journal of Computational and Nonlinear Dynamics, 2, 249-266, 2007.

[24] Kline K. A., Dynamic analysis using a reduced basis of exact modes and Ritz vectors, AIAA Journal, 24 (12), 2022-2029, 1986

[25] Balmès E., Optimal Ritz vectors for component mode synthesis using the singular value decomposition, AIAA Journal, 34 (6), 1256-1260, 1996.

[26] Wilson E. L., Yuan M. W. and Dickens J. M., Dynamic analysis by direct superposition of Ritz vectors, Earthquake Engineering and Structural dynamic, 10, 813-821, 1982.

[27] Lèger P., Application of load dependent vectors bases for dynamic substructure analysis, AIAA Journal, 28 (1), 177-179, 1990.

[28] Butcher E. A., Clearance effects on bilinear normal mode frequencies, J. Sound & Vibration, 224, 305-328, 1999.

[29] AL-Shudeifat M. A., Butcher E. A. and Burton T. D., Comparison of order reduction methodologies and identification of NNMs in structural dynamic systems with isolated nonlinearities, Proc. of IMAC-XXVII, Orlando, FL, Feb. 2009.

[30] Belendez A., Hernandez A., Marquez A. and Neip C., Analytical approximation for the period of a nonlinear pendulum, Eur. J. Phys., 27, 539-551, 2006.

Proceedings of the IMAC-XXVIII
February 1–4, 2010, Jacksonville, Florida USA


# Reduction methods for MEMS nonlinear dynamic analysis

**Paolo Tiso**
Precision and Microsystems Engineering Department
Faculty of Mechanical, Maritime and Materials Engineering
Delft University of Technology, Mekelweg 2, 2628CD Delft, The Netherlands
p.tiso@tudelft.nl

**Daniel J. Rixen**
Department of Precision and Microsystems Engineering
Faculty of Mechanical, Maritime and Materials Engineering
Delft University of Technology, Mekelweg 2, 2628CD Delft, The Netherlands
d.j.rixen@tudelft.nl

## ABSTRACT

Practical MEMS applications feature non-linear effects that are important to be realistically simulated. This typically involves large dynamic non-linear finite element (FE) models, and therefore efficient model reduction techniques are of great need. Proper Orthogonal Decomposition (POD) is a well-known technique for the effective order reduction of large dynamic (non-linear) systems. POD does not require any knowledge of the system at hand but features the disadvantage of the need of running a full simulation to extract the reduction basis. On the other hand, a basis constituted by few vibration modes enriched with modal derivatives (MD) can describe the main effect of nonlinearity without the need of a full model solution. We present a comparison of the two described reduction methods (POD and MD) applied to a geometircally non-linear micro-beam subjected to electrostatic forces.

*Keywords*: MEMS, non-linear dynamics, finite element, model order reduction, proper orthogonal decomposition, modal derivatives

## NOMENCLATURE

| | | |
|---|---|---|
| $x$ | : | micro-beam axial coordinate |
| $L$ | : | micro-beam length |
| $b$ | : | micro-beam width |
| $h$ | : | micro-beam thickness |
| $d_0$ | : | initial distance between electrodes |
| $A$ | : | micro-beam cross section area |
| $I$ | : | micro-beam bending moment of inertia |
| $E$ | : | Young modulus |
| $\rho$ | : | material density |
| $u$ | : | axial displacement |
| $w$ | : | transversal displacement |
| $\epsilon_0$ | : | dielectric constant |
| $\varepsilon$ | : | axial strain |
| $\chi$ | : | micro-beam curvature |
| $V$ | : | voltage |

| | |
|---|---|
| $\mathbf{M}$ | mass matrix |
| $\mathbf{K}$ | tangential stiffness matrix |
| $\mathbf{K}_L$ | electrostatic stiffness matrix |
| $\mathbf{g}$ | internal elastic forces |
| $\mathbf{p}$ | electrostatic forces |
| $\mathbf{u}$ | displacement vector |
| $\mathbf{q}$ | modal coordinate vector |
| $\mathbf{\Psi}$ | reduction basis |
| $\mathbf{\Phi}_i$ | generic basis vector $i$ |
| $\mathbf{U}$ | POD snapshots matrix |
| $\mathbf{C}$ | POD snapshots covariance matrix |

## 1 INTRODUCTION

Micro Electro-Mechanical Systems (MEMS) are functional machine systems with components measured in micrometers. They find promising applications in a broad range of engineering field, namely biomedical, communications, mechanical, etc. Their behaviour is inherently multi-physical. For instance, a silicon structural element is actuated either by electrostatic or thermal forces. In several cases, nonlinear effects (i.e. structural non-linearities, configuration dependent actuation forces, radiation effects) are relevant and cannot be neglected even in the early design stage. For this reason, realistic analysis of MEMS is often based on rather large multiphysics non-linear dynamic finite element (FE) models.

However, the computational effort required by such models is often prohibitive when design iterations and *what-if* analysis need to be performed. Morevoer, the direct inclusion of large MEMS models for system level simulations is not feasible. The availabililty of model reduction techniques that effectively reduce the number of degrees of freedom while preserving the accuracy of the solution is therefore of high demand in the engineering and research community.

As remarked by the review work of B.Rudnyi [1], the status of reduction techniques for linear systems is quite mature. The developments of suitable reduction methods for non-linear system is still an open research field. Different strategies have been attemped in the past in the field of non-linear structural dynamics. As for linear model reduction, the concept is to seek the solution in a subspace of well-selected modes and project the system on such subsbace. Among other techniques, two of the most used ones to create such basis are the Modal Derivatives (MD) and the Proper Othogonal Decomposition (POD), also known as *Karhunen - Love* method.

The MD method is a natural extension of vibration modes superposition in linear dynamic analysis. A basis of linear modes (i.e. vibration modes) is enriched with the derivatives of the modes themselves with respect to the modal amplitudes. This technique was proposed by [2] and [3] and later used by [4]. The work by [5] showed that the modal derivatives can also be obtained by a perturbation method applied to the governing PDE of the system. The generating basis of first order modes can also be obtained through the so-called *Wilson vectors* [6] also referred as *load dependent vectors* or *Ritz vectors*, as shown by [3]. The load dependent vectors are calculated by an Arnoldi sequence combined with a Gram-Schmidt orthogonalization. They bear the advantage that they do not require the solution of an eigenvalue problem but only a sequence of linear problems with the same factorized matrix of coefficients. Moreover, they physically account for the shape of the applied load, and therefore are particularly suitable for those applications in which this feature does not change. The MDs account for the main effect of the nolinearity present in the system and they proved to be an effective technique for dynamic thin-walled structures problems [5]. For slender structures, they contain the non-linear membrane stretching caused by the correspondent bending dominated vibration mode. Recently, MDs found applications also in the computer graphics community [7].

Instead of calculating the basis around a certain state of the system, POD requires the calculation of one of more full solution for the formation of the reduction basis. Because of its generality and easiness of implementation, POD is well-known in many fields of applications, namely fluid-dynamics, weather forecast, biology, etc.. It essentially consists of collecting snapshots from a representative full model analysis and extracting the eigenvectors associated to the largest eigenvalue of the covariance matrix formed with the sampled state of the system. Once the basis is calculated, it can be used to project the full model. Some interesting applications of POD in structural dynamics can be found in [8],[9],[10].

As indicated by [1], POD is clearly the choice as a general reduction tool when no special knowledge of the system or access to the model formulation and implementation is available. However, the potentials of the MD method could be exploited in the case of MEMS. In this contribution, we would like to compare MD and POD for a representative MEMS featuring geometrical and electrostatic nonlinearities.

## 2 MICRO-BEAM MODEL

We consider here a micro beam. The beam is of length $L$, thickness $h$ and width $b$. It is clamped at both tips and it is hold in vacuum at a distance *d* from a fixed electrode. A Voltage $V$ is applied between the electrodes so that the beam will deflect under the action of the electrostatic force. the system is sketched in Figure 1. We denote with $u(x)$ and $w(x)$ the axial and transversal displacement of the beam electrode.

Non-linear geometrical effects are taken into account via the von-Karman moderate rotation kinematic model, see for example [11]. According to this model, the axial stretching $\varepsilon$ is caused not only by the variation of the axial displacement but also by the lateral deflection of the beam, as

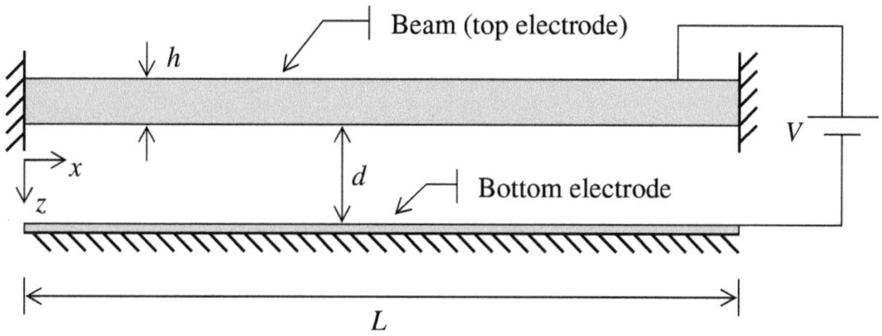


**Figure 1: the micro-beam model.**

$$\varepsilon = u_{,x} + \tfrac{1}{2} w_{,x}^2$$
$$\chi = w_{,xx} \tag{1}$$

while the bending deformation $\chi$ is defined as the curvature of the lateral deflection $w$. We denote with $(\ )_{,x}$ the derivative with respect to the $x$ coordinate.

In order to derive the nonlinear equilibrium equations, we write the energy contributions of the system, namely the kinetic energy $T$, the mechanical energy $U$ and the electrostatic energy $W$. The mechanical energy $U$ is

$$U = \frac{1}{2} \int_0^L \left[ EI\chi^2 + EA\varepsilon^2 \right] dx \tag{2}$$

where $E$ is the Young modulus and $A = bh$ and $I = \frac{1}{12} bh^3$ are the area and the moment of inertia of the beam cross section.

The electrostatic force per unit length $p(x)$ acting on the beam electrode is approximated by

$$p(x) = \frac{1}{2} \frac{\epsilon_0 b V^2}{(d - w)^2} \tag{3}$$

This form of the electrostatic forces neglects any fringing effect that occurs when the electrode deforms thus curving the electrostatic potential lines. The potential $W$ associated to the electrostatic forces is

$$W = -\frac{1}{2} \epsilon_0 V^2 \int_0^L \frac{b}{(d - w)} dx \tag{4}$$

The kinetic energy $T$ is written as

$$T = \frac{1}{2} \rho \int_0^L bh \left( \dot{u}^2 + \dot{w}^2 \right) dx \tag{5}$$

where $\rho$ is the material density. We introduce the following non-dimensional parameters:

$$
\begin{aligned}
&x = \tilde{x}L;\ w = \tilde{w}d;\ u = \tilde{u}\tfrac{d^2}{L};\ b = \tilde{b}b_0;\ h = \tilde{h}h_0;\ \tilde{I} = \tfrac{1}{12}\tilde{b}\tilde{h}^3;\ (.)' = \tfrac{(.)}{d\tilde{x}} \\
&\eta = \tfrac{m_0 L^4}{2I_0 E};\ \mu = \tfrac{md^2L^2}{I_0 E};\ \lambda = \tfrac{\epsilon_0 V^2 b_0 L^4}{2EI_0 d^3};\ \alpha = \tfrac{d^2}{h_0^2}
\end{aligned}
\tag{6}
$$

By applying the Lagrange equations to the potential of the system and using 6 we obtain the following governing equations for the micro-beam:

$$
\begin{aligned}
&\mu\ddot{u} + \alpha P' = 0 \\
&\eta\ddot{w} + (EIw'')'' + \alpha P w'' - \tfrac{\lambda b}{(1-w)^2} = 0
\end{aligned}
\tag{7}
$$

where $P = A\varepsilon$. The $(\tilde{.})$ symbol has been removed for clarity.

The details of the FE implementation are reported in the appendix. Once the PDE 7 are FE discretized, the micro-beam problem is governed by the nonlinear system of time-ODE:

$$
\begin{cases}
\mathbf{M}\ddot{\mathbf{u}}(t) + \mathbf{g}(\mathbf{u}) - \lambda\mathbf{p}(\mathbf{u}) = \mathbf{0} \\
\mathbf{u}(0) = \mathbf{u}_0 \\
\dot{\mathbf{u}}(0) = \dot{\mathbf{u}}_0
\end{cases}
\tag{8}
$$

where $\mathbf{q}$ is the displacement vector, $\mathbf{M}$ is the (configuration-independent) mass matrix, and $\mathbf{g}$ and $\mathbf{p}$ are the vectors of the nonlinear elastic and electrostatic forces, respectively. The initial displacement and velocities and written as $\mathbf{u}_0$ and $\dot{\mathbf{u}}_0$.

## 3 MODEL REDUCTION

In practical applications, the system of $N$ equations 8 is usually large. The number of unknowns can be reduced to $M$, with $M << N$, by projecting the displacement field $\mathbf{u}$ on a suitable basis of time-independent vectors $\mathbf{\Phi}$, as:

$$
\mathbf{u} = \mathbf{\Psi}\mathbf{q} \tag{9}
$$

The governing equations can then be projected on the chosen basis $\mathbf{\Psi}$ in order to make the residual orthogonal to the subspace in which the solution $\mathbf{q}$ is sought. This results in a reduced system of $M$ non-linear equations:

$$
\mathbf{\Psi}^T\mathbf{M}\mathbf{\Psi}\ddot{\mathbf{q}}(t) + \mathbf{\Psi}^T\mathbf{g}(\mathbf{\Psi}\mathbf{q}) - \lambda\mathbf{\Psi}^T\mathbf{p}(\mathbf{\Psi}\mathbf{q}) = \mathbf{\Psi}^T\mathbf{0} \tag{10}
$$

The internal force calculation is still based on operations at element level. This is expensive and dominates the time step computational efforts. For linear elastic mateiral and Kirchhoff kinematic law as the one considered in this work, the internal forces are a third order polinomial function of the nodal displacements. Once the basis $\mathbf{\Psi}$ has been chosen, it is possible to condense the polynomial factors direclty on the modal coordinates thus obtaining a reduced internal forces vector directly, see for example [7]. A similar strategy should be devised for the electrostatic force vector, which unfortunately does not feature the same polinomial structure. We refer to the numerical solution of the full model as the *full* solution, while the solution of the reduced model will be called *reduced* solution.

The key of a good reduction method is to find a suitable basis $\mathbf{\Psi}$ that is able to reproduce the full solution with a good, hopefully controlled, accuracy. In the next sections briefly describe two well-known methods for the construction of the basis for model reduction, namely the Modal Derivatives (MD) and Proper Orthogonal Decomposition (POD). We will then attempt to highlight their relative performances with respect to the application at hand.

## 4 MODAL DERIVATIVES

The projection of the governing equations on a reduction basis formed by a set of vibration modes (VM) is a well-known technique for linear structural dynamics. The main advantage of this technique is that the resulting reduced model consists of a system of uncoupled equations that can therefore be solved separately. As discussed in the introduction, several attempts has been made to extend the VM projection for nonlinear analysis. The main limitation of such approach lies in the fact that the vibration basis changes as the configuration of the system changes. It is therefore required to upgrade the basis during the numerical time integration to account for the effect of the nonlinearity.

For MEMS applications as the one considered in this contribution, the system is usually characterized by a high length-to-thickness ratio. This feature brings in significant nonlinear bending-stretching coupling effects that are usually not captured by a reduction basis formed by linear vibration modes only.

Since the VMs change with respect to the configuration, a natural way of accounting for the main effect of nonlinearity is to compute their derivatives with respect to the configuration around the reference state $(\mathbf{u}_0, \lambda_0)$ at which the modes are calculated. The direction for the derivatives are provided by the VMs chosen for the linear analysis. We want to compute:

$$\boldsymbol{\Phi}_{ij} = \frac{\partial \boldsymbol{\Phi}_i}{\partial q_j} \tag{11}$$

In other words, we would like to know how a certain mode $\boldsymbol{\Phi}_i$ changes if the structure is displaced according to the shape described by mode $\boldsymbol{\Phi}_j$. The eigenvalue problem to be solved is

$$\left[\mathbf{K}(\mathbf{u}_0) - \lambda_0 \mathbf{K}_L(\mathbf{u}_0) - \omega_i^2 \mathbf{M}\right] \boldsymbol{\Phi}_i = \mathbf{0}, \quad i, 1, 2, \dots, M \tag{12}$$

The modal derivatives can be found numerically via the finite difference method, as

$$\boldsymbol{\Phi}_{ij} \approx \frac{\boldsymbol{\Phi}_i(\mathbf{u}_0 + \boldsymbol{\Phi}_j q_j) - \boldsymbol{\Phi}_i(\mathbf{u}_0)}{q_j} \tag{13}$$

where $\boldsymbol{\Phi}_i(\mathbf{u}_0 + \boldsymbol{\Phi}_j q_j)$ denotes the modes obtained as a solution of

$$\left[\mathbf{K}(\mathbf{u}_0 + \boldsymbol{\Phi}_j q_j) - \lambda \mathbf{K}_L(\mathbf{u}_0 + \boldsymbol{\Phi}_j q_j) - \omega_i^2 \mathbf{M}\right] \boldsymbol{\Phi}_i = \mathbf{0}, \quad i, 1, 2, \dots, M \tag{14}$$

while $\boldsymbol{\Phi}_i(\mathbf{u}_0, \lambda_0)$ is the solution of the eigenvalue problem 12. This procedure is expensive, since it requires the solution of several eigenvalue problems. oreover, it is sometimes difficult to track the correspondence between the modes. The analytical approach can be used instead. A way to proceed is to differentiate the eigenvalue problem 12 with respect to the modal amplitudes. Noting that **M** does not depend nor on the displacements neither on the voltage, we can write:

$$\left[\mathbf{K} - \lambda \mathbf{K}_L - \omega_i^2 \mathbf{M}\right] \frac{\partial \boldsymbol{\Phi}_i}{\partial q_j} + \left[\frac{\partial \mathbf{K}}{\partial q_j} - \lambda \frac{\partial \mathbf{K}_L}{\partial q_j} - \frac{\partial \omega_i^2}{\partial q_j} \mathbf{M}\right] \boldsymbol{\Phi}_i = \mathbf{0} \tag{15}$$

It has been shown by [4] that a multiplication of all the mass terms by a factor $\zeta$ decreases the modal derivatives by the same factor $\zeta$. Therefore, a scaling of the mass term does not influence the mode shapes, i.e. the ojbects we are actually interested in. They proposed, as already suggested by [2], to neglect the inertia terms. Numerical experiments have shown that this approximation doesn't actually change the results. By doing so, the problem 15 becomes:

$$\left[\mathbf{K} - \lambda \mathbf{K}_L\right] \frac{\partial \boldsymbol{\Phi}_i}{\partial q_j} = \left[\lambda \frac{\partial \mathbf{K}_L}{\partial q_j} - \frac{\partial \mathbf{K}}{\partial q_j}\right] \boldsymbol{\Phi}_i \tag{16}$$

Once the $K$ vibration modes have been calculated, the reduction basis $\boldsymbol{\Psi}$ can be formed as

$$\boldsymbol{\Psi} = [\boldsymbol{\Phi}_i \; \boldsymbol{\Phi}_{ij}] \;, \; i, j = 1, \ldots, K \tag{17}$$

It can be shown that the modal derivatives are symmetric, i.e. $\boldsymbol{\Phi}_{ij} = \boldsymbol{\Phi}_{ji}$. Therefore, given $K$ vibration modes, $R = K(K+1)/2$ MD can be calculated. The reduction basis $\boldsymbol{\Psi}$ is formed by $M = K + R$ modes.

## 5 PROPER ORTHOGONAL DECOMPOSITION

As briefly introduced before, POD is a reduction technique that builds a projection basis out of a statistical elaboration of a representative solution of the full model. Let us consider a dynamic solution $\mathbf{u}(t)$ of equation 8. The method consists of collecting snaphots $\mathbf{u}^i, \; i = 1, 2, \ldots, S, \; S < N$ within the spanned time interval. The set of snapshots is collected in the matrix $\mathbf{U}$ after the average $\mathbf{u}$ is subtracted from each contribution.

$$\bar{\mathbf{u}} = \frac{1}{S} \sum_{i=1}^{S} \mathbf{u}^i \tag{18}$$

$$\overbrace{\mathbf{U}}^{N \times S} = \left[ \bar{\mathbf{u}}^1 \; \bar{\mathbf{u}}^2 \; \ldots \; \bar{\mathbf{u}}^S \right]; \; \bar{\mathbf{u}}^j = \mathbf{u}^j - \bar{\mathbf{u}} \tag{19}$$

The covariance matrix $\mathbf{C}$ is then formed as

$$\mathbf{C} = \mathbf{U}\mathbf{U}^T; \tag{20}$$

If an approximation of the snapshots set $\mathbf{u}^i$ is sought as

$$\tilde{\mathbf{u}} = \mathbf{v}_0 + \sum_{i=1}^{M} h_i(\mathbf{u} - \bar{\mathbf{u}}) \mathbf{v}^i \tag{21}$$

where $M < N$, it can be shown that the expectation

$$e \left( \|\mathbf{u} - \tilde{\mathbf{u}}\|^2 \right) \tag{22}$$

is minimized by choosing $\mathbf{v}_0 = \bar{\mathbf{u}}$, $\mathbf{v}^i = \boldsymbol{\Phi}^i$ and $h_i(\mathbf{u} - \bar{\mathbf{u}}) = (\mathbf{u} - \bar{\mathbf{u}})^T \boldsymbol{\Phi}^i$.

where $\boldsymbol{\Phi}^i$ are the solution of the eigenproblem

$$\mathbf{C}\boldsymbol{\Phi} = \lambda \boldsymbol{\Phi} \tag{23}$$

corresponding to the eigenvalues $\lambda_1 > \lambda_2 > \cdots > \lambda_N$.

It can be shown that the error $e$ is bounded as:

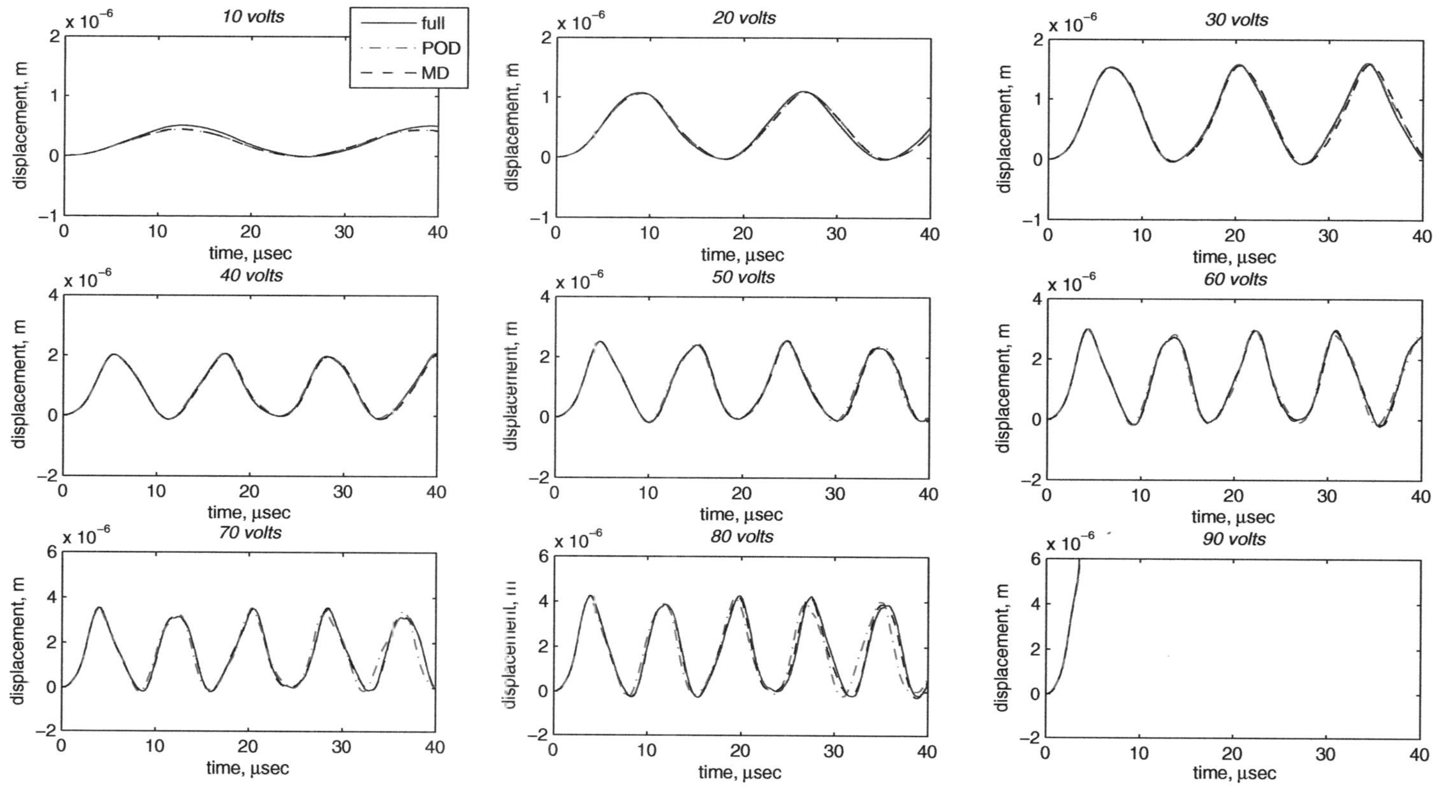


**Figure 3: Dynamic step load response at different voltages. The POD reduction basis ($M = 5$) is generated for the analysis at 50 Volts. The MD reduction basis is obtained from the eigenvalue problem for free vibrations around the static solution corresponding to 50 Volts. In this case, $M = 2$ (first two vibration modes) and $R = N(N+1)/2 = 3$. The agreement with the full solution is very good for the full range of applied voltages. The MD basis leads to better results for the higher voltage range.**

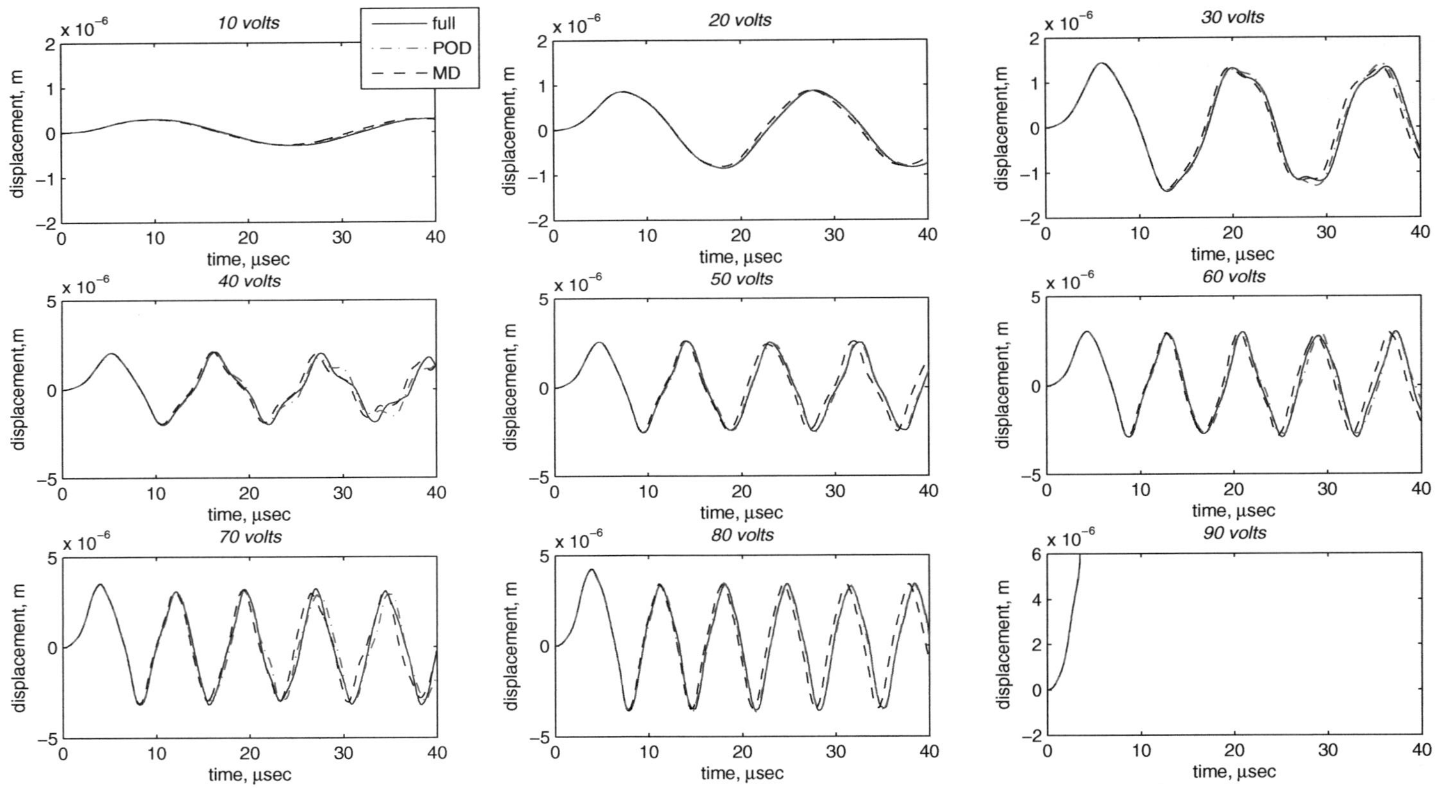


Figure 4: Dynamic square impulse load response at different voltages. The POD reduction basis ($M = 8$) is generated for the analysis at 50 Volts for the step load. The MD reduction basis is obtained from the eigenvalue problem for free vibrations around the static solution corresponding to 50 Volts. In this case, $M = 4$ (first four vibration modes) and $R = 4$, relative to the modal derivatives $\Phi_{1j},\ j = 1, 2, 3, 4$. The performance of the POD base is in this case slighly better than the MD base, most probably because of the incompleteness of the MD set.

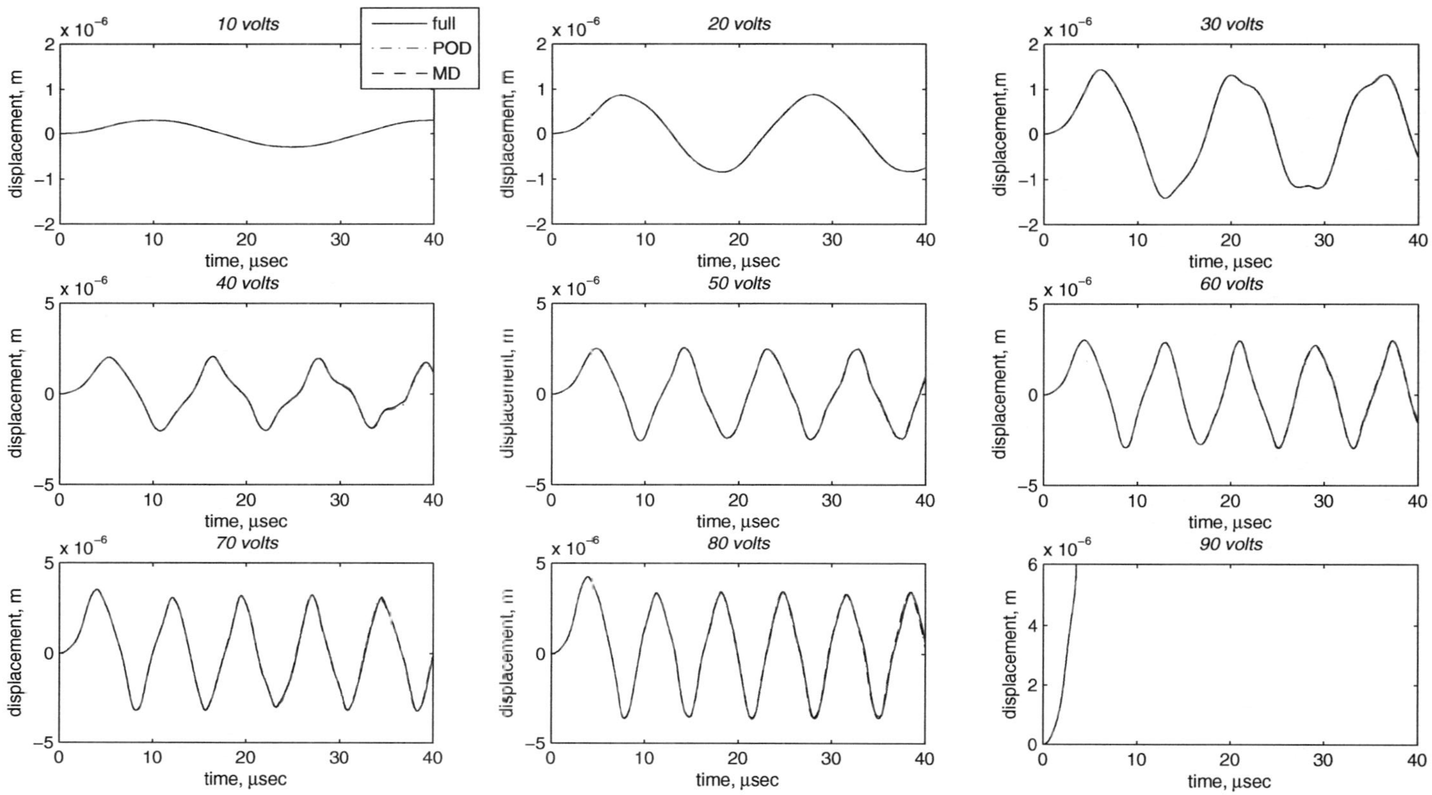


**Figure 5: Dynamic square impulse load response at different voltages. The POD reduction basis ($M = 11$) is generated for the analysis at 50 Volts for the step load. The MD reduction basis is obtained from the eigenvalue problem for free vibrations around the static solution corresponding to 50 Volts. In this case, $K = 4$ (first four vibration modes) and $R = 7$, relative to the modal derivatives $\Phi_{ij},\ i = 1, 2\ j = 1, 2, 3, 4$. The enriching of the MD basis leads to excellent accuracy.**

## 7 CONCLUSIONS

We presented a comparison between two well-known model reduction techniques, namely Proper Orthogonal Decomposition (POD) and Modal Derivatives (MD), applied to a electrostatic micro-beam featuring non-linear geometrical effects. The governing PDEs have been discretized with FE and the resulting time ODEs have been reduced with the mentioned techniques. The goal of the contribution has been to show the effectiveness of POD and MD for a range of applied dynamic voltages up to dynamic pull-in. Step load and square impulse load have been considered. The POD reduction basis has been generated from a single dynamic analysis at a voltage level (50 Volts) approximatively half of the static pull-in voltage, for a step load. Likewise, the vibration modes and the associated MDs have been calculated at the static configuration corresponding to 50 Volts. In all the cases, the POD and the MD basis where characterized by the same number of vectors for a fair comparison. Both the two methods yield excellent results for the considered problem, see Figure 3. For the case of the step load, the MD basis seems to performed better for high voltage levels. This is confirmed by the better accuracy of the MD basis in predicting the dynamic pull-in voltage, see Table 1. For the case of square pulse voltage, more vectors are required for both methods to reach a good accuracy, see Figure 5. For this load case, a POD base performs better when compared to an incomplete MD basis, see Figure 4.

The problem presented here is somewhat simplified; nevertheless, it contains all the essential non-linear features of a more realistic model. On such application, The MD method proves to be a promising reduction method as compared to the POD. MD does not require a full analysis and yields basis vectors with a clear physical interpretation, namely the main effect of non-linearity around the configuration at which the vibration modes are computed. Further work needs to be performed to find an automatic selection criterion for the most relevant MDs.

## REFERENCES


[1] **Rudnyi, E. B.** and **Korvink, J. G.**, *Review: Automatic Model Reduction for Transient Simulation of MEMS-based Devices*, Sensors Update, Vol. 11, No. 1, pp. 1432–2404, 2002.

[2] **Idelsohn, S. R.** and **Cardona, A.**, *A reduction method for nonlinear structural dynamic analysis*, Computer Methods in Applied Mechanics and Engineering, Vol. 49, pp. 253–279, 1985.

[3] **Idelsohn, S. R.** and **Cardona, A.**, *A load-dipendent basis for reduced nonlinear structural dynamics*, Computer & Structures, Vol. 20, pp. 203–210, 1985.

[4] **Slaats, P. M. A., de Jong, J.** and **Sauren, A. A. H. J.**, *Model reduction tools for nonlinear structural dynamics*, Computer & Structures, Vol. 54, pp. 1155–1171, 1995.

[5] **Tiso, P., Jansen, E.** and **M.M.Abdalla**, *A Reduction Method for Finite Element Nonlinear Dynamic Analysis of Shells, 47th AIAA/ASME/ASCE/AHS/ASC Structures, Structural Dynamics, and Materials Conference*, 2006.

[6] **Wilson, E., Juan, M.** and **Dickens, J.**, *Dynamic analysis by direct superposition of Ritz vectors*, Earthquake Engineering and Structural Dynamics, Vol. 10, pp. 813–821, 1982.

[7] **Barbič, J.** and **James, D. L.**, *Real-Time subspace integration for St. Venant-Kirchhoff deformable models*, ACM Trans. Graph., Vol. 24, No. 3, pp. 982–990, 2005.

[8] **Krysl, P., Lall, S.** and **Marsden, J. E.**, *Dimensional model reduction in non-linear finite element dynamics of solids and structures*, To appear, International Journal for Numerical Methods in Engineering.

[9] **Meyer, M.** and **Matthies, H. G.**, *Efficient model reduction in non-linear dynamics using the Karhunen-Love expansion and dual-weighted-residual methods*, Computational Mechanics, Vol. 31, No. 1, pp. 179– 191, 2003.

[10] **Zilian, A., Dinkler, D.** and **Vehre, A.**, *Projection-based reduction of fluid-structure interaction systems using monolithic space-time modes*, Computer Methods in Applied Mechanics and Engineering, Vol. 198, No. 47-48, pp. 3795 – 3805, 2009.

[11] **Brush, D. O.** and **Almroth, B. O.**, Buckling of Bars, Plates and Shells, McGraw - Hill, 1975.

[12] **Rochus, V., Rixen, D.** and **Golinval, J.-C.**, *Electrostatic coupling of MEMS structures: transient simulations and dynamic pull-in*, Nonlinear Analysis, Vol. 63, No. 5-7, pp. e1619 – e1633, 2005, invited Talks from the Fourth World Congress of Nonlinear Analysts (WCNA 2004).

[13] **Ragon, S., Gürdal, Z.** and **Watson, L.**, *A comparison of three algorithms for tracing nonlinear equilibrium paths of structural systems*, International Journal of Solids Structures, Vol. 139, pp. 689–698, 2002.

## APPENDIX: FINITE ELEMENT FORMULATION

This section briefly reports the finite element formulation for a two-dimensional Euler-Bernoulli beam element. The degrees of freedom are shown in Figure 7. We use here an iso-parametric formulation: the physical coordinate are mapped on a normalized domain via the relation:

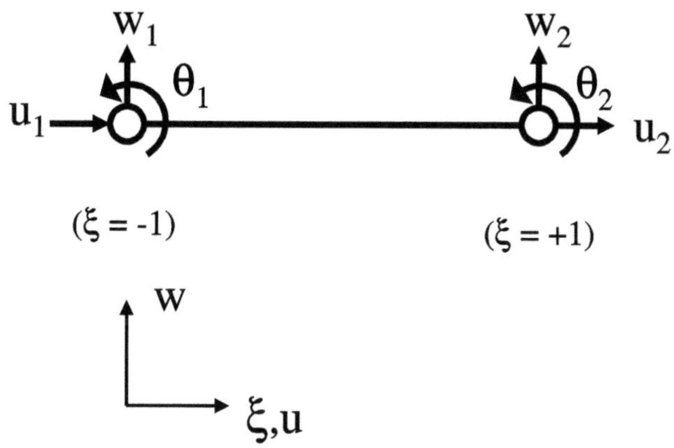


**Figure 6: Two nodes beam element.**

$$\xi(x) = \frac{2x - (x_i + x_j)}{l_i} \tag{25}$$

The displacement fields $u$ and $w$ within the element are approximated by

$$u(\xi) = \mathbf{N}_u \mathbf{u}_e \tag{26}$$

and

$$w(\xi) = \mathbf{N}_w \mathbf{w}_e \tag{27}$$

Where $\mathbf{N}_u(\xi)$ and $\mathbf{N}_w(\xi)$ is a row vector containing the Hermitian shape functions

$$\begin{aligned} \mathbf{N}_u(\xi) &= \frac{1}{2}[1-\xi \; 1+\xi] \\ \mathbf{N}_w(\xi) &= \left[\frac{1}{4}(-1+\xi)^2(2+\xi) \; \frac{l_e}{8}(-1+\xi)^2(1+\xi) \; -\frac{1}{4}(-2+\xi)(1+\xi)^2 \; \frac{l_e}{8}(-1+\xi)(1+\xi)^2\right] \end{aligned}$$

and the element nodal displacement $\mathbf{u}_e$ and $\mathbf{w}_e$ are organized as:

$$\begin{aligned} \mathbf{u}_e &= [u_1 \; u_2]^T \\ \mathbf{w}_e &= [w_1 \; \theta_1 \; w_2 \; \theta_2]^T \end{aligned}$$

The internal force vector $\mathbf{g}$ can be split in two parts, namely the axial term $\mathbf{g}_u$ and the transversal term $\mathbf{g}_w$, as:

$$\mathbf{g} = \begin{bmatrix} \mathbf{g}_u \\ \mathbf{g}_w \end{bmatrix} \tag{28}$$

the axial and transversal internal force vector are:

$$\mathbf{g}_u = \frac{1}{2}\int_{-1}^{1} P\mathbf{N}_u^{'T} d\xi \tag{29}$$

and

$$\mathbf{g}_w = \int_{-1}^{1} \left[ \frac{1}{2l_e} P(\mathbf{N}_w^{'}\mathbf{w}_e)\mathbf{N}_w^{'T} + \frac{1}{2l_e^3} M\mathbf{N}_w^{''T}\mathbf{N}_w^{''} \right] d\xi \tag{30}$$

where $l_e$ is the element length and

$$P = EA\varepsilon = EA\left(\frac{1}{l_e}\mathbf{N}_u^{'}\mathbf{u}_e + \frac{1}{2}\frac{1}{l_e^2}(\mathbf{N}_w^{'}\mathbf{w}_e)^2\right) \tag{31}$$

and

$$M = EI\chi = \frac{1}{l_e^4} EI\mathbf{N}_w^{''T}\mathbf{N}_w^{''} \tag{32}$$

The apex $()'$ stands for the derivative with respect of $\xi$. The structural stiffness matrix $\mathbf{K}$ is the derivative of the internal forces with respect to the nodal displacements. It can be partitioned into axial and transversal contribution, as

$$\mathbf{K} = \begin{bmatrix} \mathbf{K}_{uu} & \mathbf{K}_{uw} \\ \mathbf{K}_{uw}^T & \mathbf{K}_{ww} \end{bmatrix} \tag{33}$$

where the different terms are written as:

$$\mathbf{K}_{uu} = \frac{\partial \mathbf{g}_u}{\partial \mathbf{u}_e} = \frac{1}{2l_e}\int_{-1}^{1} EA\mathbf{N}_u^{'T}\mathbf{N}_u^{'} d\xi \tag{34}$$

$$\mathbf{K}_{uw} = \frac{\partial \mathbf{g}_u}{\partial \mathbf{w}_e} = \frac{1}{2l_e^2}\int_{-1}^{1} EA(\mathbf{N}_w^{'}\mathbf{w}_e)\mathbf{N}_u^{'T}\mathbf{N}_w^{'} d\xi \tag{35}$$

$$\mathbf{K}_{ww} = \frac{\partial \mathbf{g}_w}{\partial \mathbf{w}_e} = \frac{1}{2l_e^3}\int_{-1}^{1} EI\mathbf{N}_w^{''T}\mathbf{N}_w^{''} d\xi + \frac{1}{2l_e^2}\int_{-1}^{1} EA(\mathbf{N}_w^{'}\mathbf{w}_e)^2\mathbf{N}_w^{'T}\mathbf{N}_w^{'} d\xi + \frac{1}{2l_e}\int_{-1}^{1} P\mathbf{N}_w^{'T}\mathbf{N}_w^{'} d\xi \tag{36}$$

The mass matrix is found as

$$\mathbf{M}_u = \frac{l_i b_i h_i \rho}{2} \int_{-1}^{1} \mathbf{N}_u^T \mathbf{N}_u d\xi \tag{37}$$

$$\mathbf{M}_w = \frac{l_i b_i h_i \rho}{2} \int_{-1}^{1} \mathbf{N}_w^T \mathbf{N}_w d\xi \tag{38}$$

$$\mathbf{M} = \begin{bmatrix} \mathbf{M}_u & \mathbf{0} \\ \mathbf{0} & \mathbf{M}_w \end{bmatrix} \tag{39}$$

The electrostatic loads are written as

$$\mathbf{p}_e = \frac{b_e l_e}{2} \int_{-1}^{1} \frac{\mathbf{N}_w^T}{(d - \mathbf{N}_w \mathbf{w}_e)^2} d\xi \tag{40}$$

where $b_e$ is the element width. By differentiating the electrostatic load $\mathbf{p}_e$ with respect to the displacement $\mathbf{w}_e$ we obtain the electrostatic tangential stiffness matrix $\mathbf{K}_{L_e}$ as

$$\mathbf{K}_{L_e} = \frac{b_e l_e}{2} \int_{-1}^{1} \frac{\mathbf{N}_w^T \mathbf{N}_w}{(d - \mathbf{N}_w \mathbf{w}_e)^3} d\xi \tag{41}$$

The system of equations 16 requires the calculation of the derivative of the structural and electrostatic tangential stiffness matrix $\mathbf{K}$ and $\mathbf{K}_L$ with respect to the modal amplitudes $q_j$. To express this contribution, we introduce the notation

$$\mathbf{u}_e = \mathbf{L}_e \mathbf{u} = \mathbf{L}_e \mathbf{\Psi} \mathbf{q} \tag{42}$$

where the matrix $\mathbf{L}_e$ is a boolean matrix that contracts the global displacement vector $\mathbf{u}$ to the element displacements $\mathbf{u}_e$. We obtain then:

$$\frac{\partial \mathbf{K}_{uu}}{\partial q_j} = \mathbf{0} \tag{43}$$

$$\frac{\partial \mathbf{K}_{uw}}{\partial q_j} = \frac{1}{2 l_e^3} \int_{-1}^{1} EA(\mathbf{N}_w^{'} \mathbf{w}_e) \mathbf{N}_e^{'T} \mathbf{N}_w^{'} (\mathbf{N}_w^{'T} \mathbf{L}_e \mathbf{\Psi}_j) d\xi \tag{44}$$

$$\frac{\partial \mathbf{K}_{ww}}{\partial q_j} = \frac{1}{2 l_e^3} \int_{-1}^{1} 3EA(\mathbf{N}_w^{'} \mathbf{w}_e) \mathbf{N}_w^{'T} \mathbf{N}_w^{'} (\mathbf{N}_w^{'} \mathbf{L}_e \mathbf{\Psi}_j) d\xi \tag{45}$$

$$\frac{\partial \mathbf{K}_L}{\partial q_j} = \frac{3 l_e b_e}{2} \int_{-1}^{1} \frac{\mathbf{N}_w^T \mathbf{N}_w (\mathbf{N}_w \mathbf{L}_e \mathbf{\Psi}_j)}{(d - \mathbf{N}_w \mathbf{w}_e)^4} d\xi \tag{46}$$

Proceedings of the IMAC-XXVIII
February 1–4, 2010, Jacksonville, Florida USA


# On the Identification of Hysteretic Systems, Part I: an Extended Evolutionary Scheme

**Keith Worden** and **Graeme Manson**

*Dynamics Research Group*
*Department of Mechanical Engineering*
*University of Sheffield*
*Mappin Street, Sheffield S1 3JD, UK*
*k.worden@sheffield.ac.uk*

**Abstract:** Evolutionary optimisation schemes have recently established themselves as an effective means of identifying dynamical systems, even when the parameter estimation problem is complicated by nonlinearity-in-the-parameters and the presence of unmeasured states. In particular, such an approach to the parameter estimation problem for hysteretic systems has proved rather successful. Previous work by the authors has adopted the Differential Evolution (DE) algorithm of Storn and Price as the evolutionary algorithm of choice for the identification problem. Although the algorithm has proved very effective in the identification context, a minor disadvantage manifests itself in the need to set algorithm hyperparameters for the optimisation. The objective of the current paper is simply to present a recently-developed variant of the DE algorithm – the Self-Adaptive Differential Evolution algorithm (SADE) – which learns and adapts a subset of its own hyperparameters throughout the optimisation process. The use of the algorithm for the hysteretic system identification problem is illustrated using data from a computer model and it is shown that the algorithm provides several orders-of-magntitude improvement on the minimisation of the problem objective function.

**Keywords:** Hysteresis, the Bouc-Wen model, system identification, self-adaptive differential evolution.

## 1. Introduction

Hysteretic or memory-dependent phenomena are observed in many areas of physics and engineering, such as: electricity and magnetism, material phase transitions and the elasto-plasticity of solids [1]. In mechanical vibrations, the elasto-plasticity of some vibrating components introduces nonlinearities which can be regarded as of hysteretic type. Of the many available models of hysteresis, one which proves versatile enough to accurately describe many random vibrations of the above type, is the Bouc-Wen model [2,3]. The literature on the Bouc-Wen model is extensive and no attempt is made here to provide comprehensive references; instead the reader is referred to the recent book [4], which provides an excellent summary of research on the model and can be consulted as a comprehensive guide to the literature. One of the main problems associated with adopting a Bouc-Wen model as a possible model of a physical system, is the identification of the model parameters. Fairly recent work has shown that evolutionary optimisation schemes (genetic algorithms and differential evolution) offer effective means of identifying dynamical systems, even when the parameter estimation problem is complicated by nonlinearity and the presence of unmeasured states, as in the case of the Bouc-Wen system [5,6].

This paper forms the first in a short sequence considering the system identification (SI) problem for hysteretic systems. The basic model for parameter estimation will be assumed to be the Bouc-Wen model, as this has proved particularly versatile in the past; however, the approach is quite general and would accommodate many forms of parametric models. As indicated above, one successful approach to the SI problem in the past has been to cast it in the form of an optimisation problem and then to apply an evolutionary algorithm. Because of its real-valued nature, the Differential Evolution (DE) algorithm [7], has proved to be particularly effective. There are various forms of the DE algorithm and each of these requires the specification of a number of hyperparameters;

however, one recent development in evolutionary programming has been the extension of the DE algorithm to produce the *self-adaptive* (SA) variant – SADE – which is able to switch between versions of the algorithm as appropriate and to adapt its hyperparameters as needed [8]. The objective of the current paper is to demonstrate a parameter estimation scheme for Bouc-Wen systems based on an implementation of SADE. The results are compared with the standard DE approach using data from a simulated hysteretic system.

The layout of the paper is as follows: Section Two briefly describes the Bouc-Wen model of hysteresis considered here. Section Three outlines the standard Differential Evolution (DE) approach to parameter estimation and presents some estimates for a simulated problem. Section Four introduces the SADE algorithm and shows how improved estimates for the identification problem arise from its use. Section Five is concerned with the conclusions from the study.

## 2. The Bouc-Wen Hysteresis Model

The general Bouc-Wen model [2,3] is a nonlinear hysteretic restoring force model where the total restoring force $Q(y,\dot{y})$ can be composed of a polynomial non-hysteretic and a hysteretic component based on the displacement $y$, and velocity $\dot{y}$, time-histories. The general Single-Degree-of-Freedom (SDOF) hysteretic system described in the terms of Wen [3], is represented below where $g(y,\dot{y})$ is the polynomial part of the restoring force and $z(y,\dot{y})$ the hysteretic,

$$m\ddot{y} + g(y,\dot{y}) + z(y,\dot{y}) = x(t) \tag{1}$$

$m$ is the mass of the system and $x(t)$ is the excitation force. For the purposes of this paper, the polynomial part of the restoring force will be assumed linear, so that.

$$m\ddot{y} + c\dot{y} + ky + z(y,\dot{y}) = x(t) \tag{2}$$

The hysteretic component is defined by Wen [3] via the additional equation of motion,

$$\begin{aligned} \dot{z} &= -\alpha\,|\,\dot{y}\,|\,z^{n} - \beta\dot{y}\,|\,z^{n}\,| + A\dot{y} && \text{for } n \text{ odd} \\ \dot{z} &= -\alpha\,|\,\dot{y}\,|\,z^{n-1}\,|\,z\,| - \beta\dot{y}z^{n} + A\dot{y} && \text{for } n \text{ even} \end{aligned} \tag{3}$$

The parameters $\alpha$, $\beta$ and *n* govern the shape and smoothness of the hysteresis loop. The equations offer a simplification from the point of view of parameter estimation in that the stiffness term in (2) can be combined with the $A\dot{y}$ term in the state equation for $z$. As a system identification problem, this set of equations presents a number of difficulties, foremost are:

- The variables available from measurement will generally be the input *x* and some form of response; in this paper the response variable will be assumed to be displacement *y*, although the identification problem here can just as easily be formulated in terms of velocity or acceleration. The state *z* is not measurable and therefore it is not possible to use equation (3) directly in a least-squares formulation.
- The parameter *n* enters the state equation (3) in a nonlinear way; this means that a linear least-squares approach is not applicable to the estimation of the full parameter set, some iterative nonlinear least squares approach is needed as in [9].

How these problems are addressed in the context of the current study is discussed in the following section.

## 3. Hysteretic System Identification using Differential Evolution

In order to benchmark the SADE algorithm, the standard DE algorithm will be used to provide a reference solution to the identification problem. For the sake of completeness, a brief overview of the basic DE algorithm will be given here; for more detail, the reader is referred to [7]. As in all evolutionary optimisation procedures, a population of possible solutions (here, the vector of parameter estimates), is iterated in such a way that succeeding generations of the population contain better solutions to the problem in accordance with the Darwinian principle of 'survival of the fittest'. The problem is framed here as a minimisation problem with the cost function defined as a normalised mean-square error between the 'measured' data and that predicted using a given parameter estimate, i.e.,

$$J(m,c,\alpha,\beta,A)=\frac{100}{N\sigma_y^2}\sum_{i=1}^{N}(y_i-\hat{y}_i(m,c,\alpha,\beta,A))^2 \tag{4}$$

where $\sigma_y^2$ is the variance of the 'measured' sequence of displacements $\{y_i, i=1,\ldots,N\}$ and the caret denotes a predicted quantity. This cost function has the following useful property; if the mean of the output signal $\overline{y}$ is used as the model i.e. $\hat{y}_i=\overline{y}$ for all *i*, the cost function is 100.0, i.e.,

$$J(m,c,\alpha,\beta,A)=\frac{100}{N\sigma_y^2}\sum_{i=1}^{N}(y_i-\overline{y})^2=100 \tag{5}$$

Previous experience has shown that a cost value of less than 5.0 represents a good model or parameter estimate, while one with less than 1.0 can usually be considered excellent. Note that this definition of cost function could quite easily be used with velocity or acceleration data; this means that whatever data is sampled, there will be no need to apply numerical differentiation or integration procedures. A further advantage of this approach is that the optimisation does not need measurements of *z*; the correct prediction for the state is implicit in the approach. This overcomes the first of the problems discussed in the last section.

The data for the current study were generated by simulation. The coupled equations (2) and (3) were integrated forward in time in Matlab [10] using a fixed-step fourth-order Runge-Kutta scheme for initial value problems [11]. A better solution could potentially be found by using an adaptive solver like the $(4,5)^{th}$-order Runge-Kutta method for the solution of non-stiff problems encapsulated in the Matlab function ode45; it is adaptive and can adjust the step-size in order to obtain a prescribed error tolerance. However, it was shown in [12] that the use of the adaptive scheme in the context of evolutionary system identification can lead to strange results. The parameters for the baseline system adopted here were: $m=1$, $c=20$, $k=0$, $\alpha=1.5$, $\beta=-1.5$, $n=2$ and $A=6680$. The excitation $x(t)$ was a Gaussian random sequence with mean zero and standard deviation 9.92. These are the same parameter values as those adopted in [6]. The step-size (or sampling interval) was taken as 0.004 s, corresponding to a sampling frequency of 250 Hz. As the effect of noise was considered in [6], the current study uses uncorrupted data. The 'training set' or identification set used here was composed of 1000 points corresponding to a record duration of 4 seconds.

Once the data were generated, differential evolution was applied to the identification problem using a parameter vector $(m,c,\alpha,\beta,A)$. The parameter *n* was not included; the reason being that if *n* is constrained to be small integer (1 to 4, say), it is arguably best selected by a cross-validation procedure using a validation set as commonly applied by the machine learning community [13]. It is conceivable that a non-integer value of *n* could prove useful under certain circumstances; however, including *n* in the parameter vector as a real parameter raises the problem of fractional powers generating complex data.

The standard DE algorithm of reference [8] attempts to transform a randomly generated initial population of parameter vectors into an optimal solution through repeated cycles of evolutionary operations, in this case: mutation, crossover and selection. In order to assess the suitability of a certain solution, an objective function relating to the value of this solution must be constructed; in the case of minimisation problems, it is referred to as a cost function and the cost appropriate to the identification problem here is the one given above in equation (4).

Figure 1 shows a schematic for the procedure for evolving between populations. The following process is repeated with each vector within the current population being taken as the *target vector*; each of these vectors has an associated cost taken from equation (4). Each target vector is pitted against a *trial vector* in a selection process which results in the vector with lowest cost advancing to the next generation. The process for constructing the trial vector involves variants of the standard evolutionary operators: mutation and crossover.

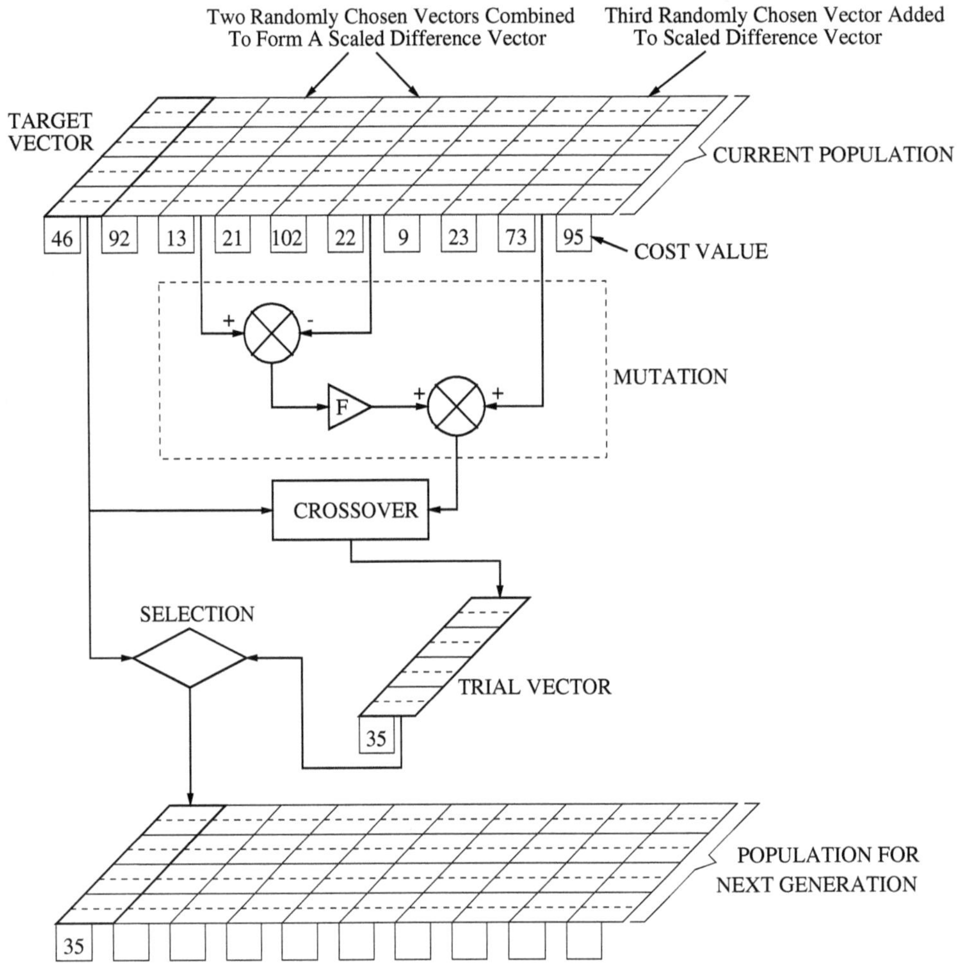


*Figure 1.* Schematic for the standard Differential Evolution algorithm.

The mutation procedure used in basic DE employs vector differentials. Two vectors $A$ and $B$ are randomly chosen from the current population to form a vector differential $A-B$. A *mutated vector* is then obtained by adding this differential, multiplied by a scaling factor, $F$, to a further randomly chosen vector $C$ to give the overall expression for the mutated vector: $C+F(A+B)$. The scaling factor, $F$, is often found have an optimal value between 0.4 and 1.0.

The *trial vector* is the child of two vectors: the target vector and the mutated vector, and is obtained via a crossover process; in this work uniform crossover is used. Uniform crossover decides which of the two parent vectors contributes to each chromosome of the trial vector by a series of $D-1$ binomial experiments. Each experiment is mediated by a crossover parameter $C_r$ (where $0 \leq C_r \leq 1$). If a random number generated from the uniform distribution on [0,1] is greater than $C_r$, the trial vector takes its parameter from the target vector, otherwise the parameter comes from the mutated vector.

This process of evolving through the generations is repeated until the population becomes dominated by only a few low cost solutions, any of which would be suitable.

In this case, the DE algorithm was initialised with a population of randomly selected parameter vectors or individuals. The parameters were generated using uniform distributions on ranges covering one order of magnitude above and below the true values as in [6]. The approach in [6] to keeping the evolving parameters within the specified ranges involved using a penalty function; in the current study, parameters leaving the ranges were simply re-initialised randomly in the allowed intervals.

A population of 30 individuals was chosen for the DE runs with a maximum number of generations of 200. In order to sample different random initial conditions for the DE algorithm, 10 independent runs were made. The other parameters chosen used for DE were an *F*-value of 0.9 and a crossover probability of 0.5, as these values had proved to be effective in a number of previous studies; this completes the specification of the DE.

| Parameter | True Value | Best Model | % Error | Minimum | Maximum | Mean | Standard Deviation |
|---|---|---|---|---|---|---|---|
| $m$ | 1.0 | 0.9995 | 0.05 | 0.996 | 1.002 | 1.000 | 0.001 |
| $c$ | 20.0 | 20.011 | 0.06 | 19.950 | 20.043 | 20.011 | 0.029 |
| $\alpha$ | 1.5 | 1.496 | 0.27 | 1.419 | 1.535 | 1.494 | 0.032 |
| $\beta$ | -1.5 | -1.423 | 5.13 | -1.683 | -1.243 | -1.501 | 0.143 |
| $A$ | 6680.0 | 6678.6 | 0.02 | 6657.8 | 6695.2 | 6676.6 | 11.3 |

*Table 1.* Summary results for 10 DE identification runs on the simulation data.

Each of the 10 runs of the DE algorithm converged to an excellent solution to the problem in the sense that cost function values of less than 0.001 were obtained in all cases; the summary results are given in Table 1. The best solution gave a cost function value $7.36\times10^{-5}$.

A comparison between the 'true' and predicted responses for the best parameter set is given in Figure 2.

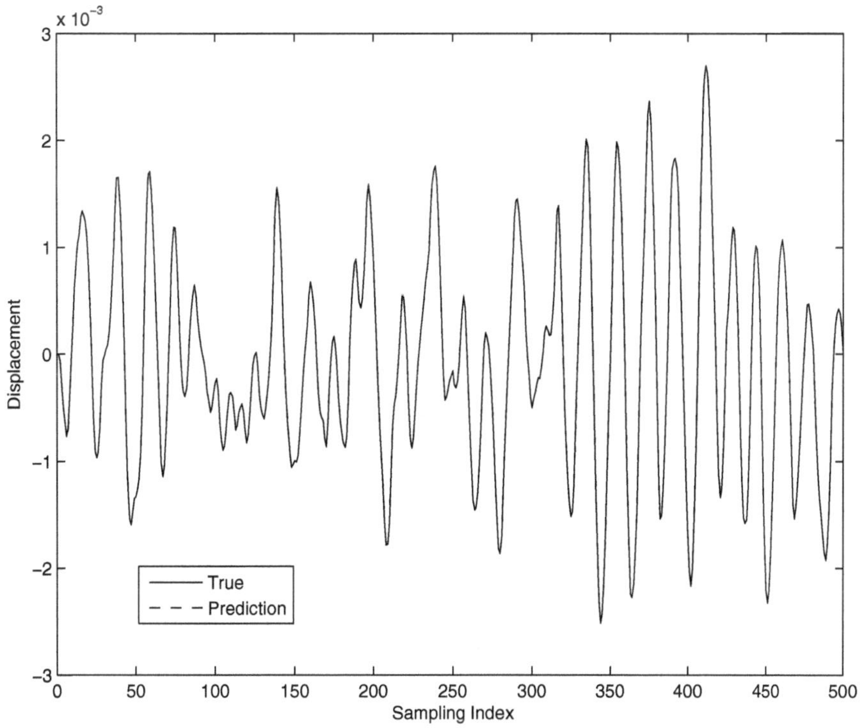


*Figure 2.* Comparison between 'true' response and prediction of best DE model.

The results are interesting, the accuracy of the estimates is consistent with that obtained in [6]. The range of results obtained over the 10 runs bracket the true results and could therefore be used as the constraints on a refinement of the parameters as used in [6]; however, this option was not pursued here. The predicted response (Figure 1) is excellent; even with the standard DE algorithm, the predictions are indistinguishable from the 'true' values of displacement.

## 4. Hysteretic System Identification using SADE

The reader will readily observe that the standard implementation of the DE algorithm described above requires the prior specification of a number of *hyperparameters*. Setting aside the requirement to define a population size, maximum number of iterations etc., the algorithm needs *a priori* values for the scaling factor $F$ and crossover probability $C_r$ in order to function. The values used above have been chosen on the basis of previous experience with the DE algorithm in which they gave good results; however, they are not guaranteed to work as well in all situations and an algorithm which establishes 'optimum' values for these parameters during the course of the evolution is clearly desirable. Such an algorithm is available in the form of the Self-Adaptive Differential Evolution (SADE) algorithm [8,14]; the description of the algorithm here follows closely to [14]. The development of the SADE algorithm begins with the observation that Storn and Price, the originators of DE, arrived at five possible strategies for the mutation operation [15]:

1. *rand1*: $M = A + F(B - C)$,
2. *best1*: $M = X^* + F(B - C)$,
3. *current-to-best*: $M = T + F(X^* - T) + F(B - C)$,
4. *best2*: $M = X^* + F(A - B) + F(C - D)$,
5. *rand2*: $M = A + F(B - C) + F(D - E)$,

where $T$ is the current trial vector, $X^*$ is the vector with (currently) best cost and $(A, B, C, D, E)$ are randomly-chosen vectors in the population distinct from $T$. $F$ is a standard (positive) scaling factor. The SADE algorithm also uses multiple variants of the mutation algorithm as above; however these are restricted to the following four:

1. *rand1*,
2. *current-to-best2*: $M = T + F(X^* - T) + F(A - B) + F(C - D)$,
3. *rand2*:
4. *current-to-rand*: $M = T + K(A - T) + F(B - C)$.

In the strategy *current-to-rand*, $K$ is defined as a coefficient of combination and would generally be assumed in the range *[-0.5, 1.5]*; however, in the implementation of [14] and the one used here, the prescription $K = F$ is used to essentially restrict the number of tunable parameters. The SADE algorithm uses the standard crossover approach, except that at least one crossover is forced in each operation on vectors. If mutation moves a parameter outside its allowed (predefined) bounds, it is pinned to the boundary. Selection is performed exactly as in DE; if the trial vector has smaller (or equal) cost to the target, it replaces the target in the next generation.

So far, this is just DE with multiple mutation strategies, adaption is yet to be defined. First of all, a set of probabilities are defined: $\{p_1, p_2, p_3, p_4\}$, which are the probabilities that a given mutation strategy will be used in forming a trial vector. These probabilities are intialised to be all equal to 0.25. When a trial vector is formed during SADE, a roulette wheel selection is used to choose the mutation strategy on the basis of the probabilities. (Initially, all mutation strategies are equally probable.) At the end of a given generation, the numbers of trial vectors successfully surviving to the next generation from each strategy are recorded as: $\{s_1, s_2, s_3, s_4\}$; the numbers of trial vectors from each strategy which are discarded are recorded as: $\{d_1, d_2, d_3, d_4\}$. At the beginning of a SADE run, the survival and discard numbers are established over the first $N_L$ generations, this interval is called the *learning period*. At the end of the learning period, the strategy probabilities are updated by,

$$p_i = \frac{s_i}{s_i + d_i} \tag{6}$$

After the learning period, the probabilities are updated every generation but using survival and discard numbers established over a moving window of the last $N_L$ generations. It is clear that the approach given here provides an adaptive means of establishing the preferred mutation strategies. As stated above, SADE also incorporates adaption or variation on the parameters or (hyperparameters) $F$ and $C_r$. The scaling factor $F$ mediates the convergence speed of the algorithm, with large values being appropriate to global search early in a run and small values being consistent with local search later in the run. The implementation of SADE used here follows largely follows [16] and differs only in one major aspect, concerning the adaption of $F$. Adaption of the parameter $C_r$ is based on accumulated experience of the successful values for the parameter over the run. It is assumed that the crossover probability for a trial is normally distributed about a mean $\overline{C_r}$ with standard deviation 0.1. At initiation, the parameter $C_r$ is set to 0.5 to give equal likelihood of each parent contributing a chromosome. The crossover probabilities are then held fixed for each population index for a certain number of generations and then resampled. In a rather similar manner to the adaption of the strategy probabilities, the $C_r$ values for trial vectors successfully passing to the next generation are recorded over a certain greater number of generations and their mean value is adopted as the next $\overline{C_r}$. The record of successful trials is cleared at this point in order to avoid long-term memory effects. Note that this prescription means that the crossover probabilities may be resampled several times within the period of learning successful values. The paper [16] differed from the current implementation in adopting random selection of $F$ for each trial from a Gaussian distribution with mean 0.5 and standard deviation 0.3; (the $F$ value was also forcibly constrained within the interval (0,2]); whereas the version here adapts $F$ in essentially the same manner as $C_r$ but uses the Gaussian $N(0.5,0.3)$ for the initial distribution.

The concerned reader may feel that the specification of the various learning periods and resampling periods simply replaces the requirement for some hyperparameters with a requirement for others; however, the advantage of the adaptive algorithm is that experiments have shown that the results are largely insensitive to the learning period etc.

The version of SADE described in [14] is slightly more complicated than that described above as it incorporates the possibility of switching to a local search procedure every now and then; in [14], the authors switch to local search using a sequential quadratic programming step every 500 generations. The algorithm in [14] also has the possibility of incorporating various optimisation constraints; the implementation used in the current study only applies constraints in the sense that the identification parameters are not allowed to leave a given range.

The use of the SADE algorithm is now illustrated on the same identification problem as that considered earlier. To show how robust the algorithm is, the results are presented for the *first* attempt with the algorithm, where the learning period was simply taken as a plausible 10 generations; updates were subsequently applied every 10 generations. As before, the algorithm used a population of 30 individuals and was allowed to run for 200 generations for 10 independent runs with different initial populations, this means that the same number of cost function evaluations took place as in the standard DE run. The results of the SADE run are given in Table 2.

| Parameter | True Value | Best Model | % Error | Minimum | Maximum | Mean | Standard Deviation |
|---|---|---|---|---|---|---|---|
| $m$ | 1.0 | 1.000 | 0.0005 | 1.000 | 1.000 | 1.000 | 0.00008 |
| $c$ | 20.0 | 20.00 | 0.0006 | 19.999 | 20.001 | 20.000 | 0.0008 |
| $\alpha$ | 1.5 | 1.500 | 0.0004 | 1.498 | 1.501 | 1.499 | 0.0008 |
| $\beta$ | -1.5 | -1.500 | 0.005 | -1.502 | -1.495 | -1.499 | 0.0022 |
| $A$ | 6680.0 | 6680.0 | 0.003 | 6679.7 | 6682.0 | 6680.2 | 0.67 |

*Table 2.* Summary results for 10 SADE identification runs on the simulation data.

The results from SADE show a radical improvement on the results from the standard DE given in Table 1. The cost function value for the best run is $1.51\times10^{-9}$, an improvement over DE of over four orders of magnitude. There is little point here in showing a comparison between the model predictions and the original data as the curves are indistinguishable.

## 5. Conclusions

The main conclusion of the paper is that the self-adaptive variant of the Differential Evolution algorithm appears to give considerably improved parameter estimates when used for the system identification of Bouc-Wen-type hysteretic systems. This conclusion must be considered in the light of two observations, the first of these is concerned with the nature of the comparison between the algorithms here. When one compares algorithms in a rigorous fashion, one should take great care that the algorithms are allowed to function at the upper limits of their capability; this usually requires the comparison of many runs taken over the allowed ranges of any algorithm parameters or hyperparameters. Alternatively, if a principled method, like cross-validation, exists for the selection of 'optimal' hyperparameters, this should be used. This approach is clearly not taken here; the comparison is made on the following basis. The parameters for standard DE – the scaling factor and crossover probability – were chosen on the basis of experience with a number of previous studies, where they proved very effective over a diverse range of problems. This gives confidence that, even if the DE algorithm was not truly operating at the upper limits of its effectiveness, it would nonetheless be expected to perform well. In contrast, the results for SADE are shown from the very first algorithm run on the problem of interest with no comparable prior experience of the setting of parameters (in the case of SADE, these were the length of the learning period and the spacing between updates). Under the circumstances, in contrast with the DE situation, one would not have comparable confidence in the performance of the algorithm, unless the benefits conferred by the adaption process are not compromised by sensitivity to the setting of the learning period etc. The results here are presented simply in the spirit of exposing a new algorithm which could very well prove to give improved results for the system identification problem; however, it is anticipated that the conclusions would only be reinforced by a truly rigorous comparison.

The second observation which affects the impact of the results presented here, concerns the absence of 'measurement' noise in the data considered. This omission was deliberate in the sense that it was intended here to investigate how well the algorithm could extract precise estimates of the system parameters in ideal conditions. In reality, measurement noise will be present in either the excitation or response (or both) and this will impose a lower limit on the cost function which can be attained. It is likely that with any realistic levels of measurement noise, the minimum cost would be attained in similar times with both the standard and adaptive algorithms and this would diminish the apparent impact of the adaptive algorithm in the identification context. A similar remark would probably limit the superiority of the adaptive algorithm when the training data is generated by an adaptive solver (or a real process) and not the fixed-step solver used here for the optimisation process.

## References

[1] Visitin, A., *Differential Models of Hysteresis*, Springer, Berlin, 1994.

[2] Bouc, R., "Forced Vibration of Mechanical System With Hysteresis", *Proceedings of 4th Conference on Nonlinear Oscillation, Prague, Czechoslovakia*, 1967.

[3] Wen, Y., "Method for Random Vibration of Hysteretic Systems", *American Society of Civil Engineers - Journal of the Engineering Mechanics Division,* XX, 1976, pp.249-263.

[4] Ikhouane, F., and Rodellar, J., *Systems with Hysteresis: Analysis, Identification and Control using the Bouc-Wen Model*, Wiley-Blackwell, 2007.

[5] Deacon, B.P. and Worden, K., "Identification of Hysteretic Systems using Genetic Algorithms", *In Proceedings of EUROMECH - 2nd European Nonlinear Oscillations Conference, Prague*, 1996, pp.55-58.

[6] Kyprianou, A., Worden, K. and Panet, M., "Identification of Hysteretic Systems using the Differential Evolution Algorithm", *Journal of Sound and Vibration*, **248**, 2001, pp.289-314.

[7] Price, K. and Storn, R., "Differential Evolution – a Simple and Efficient Heuristic for Global Optimization over Continuous Spaces", *Journal of Global Optimization*, **11**, 1997, pp.341-359.

[8] Qin, A.K. and Suganthan, P.N., "Self-Adaptive Differential Evolution Algorithm for Numerical Optimization", *in IEEE Congress on Evolutionary Computation (CEC 2005), Edinburgh, Scotland*, 2005.

[9] Yar, M. and Hammond, J.K., "Parameter Estimation for Hysteretic Systems", *Journal of Sound and Vibration*, **117**, 1987, pp.161-172.

[10] *Matlab v7*, The Mathworks, 2004.

[11] Press, W.H., Teukolsky, S.A., Vetterling, W.T. and Flannery, B.P., *Numerical Recipes 3rd Edition: The Art of Scientific Computing*, Cambridge University Press, 2007.

[12] Worden, K., "The Fitness Landscape for Parameter Estimates for Hysteretic Systems", *in Proceedings of 2nd International Conference on Engineering Dynamics – ICEDyn09, Ericiera, Portugal*, 2009.

[13] Bishop, C.M., *Neural Networks for Pattern Recognition*, Oxford University Press, 1995.

[14] Huang, V.L., Qin, A.K. and Suganthan, P.N., "Self-adaptive Differential Evolution Algorithm for Constrained Real-Parameter Optimization", *in IEEC Congress on Evolutionary Computation (CEC 2006), Vancouver, Canada*, 2006, pp.17-24.

[15] http://www.icsi.berkeley.edu/~storn/code.html: Accessed 27th October 2009.

Proceedings of the IMAC-XXVIII
February 1–4, 2010, Jacksonville, Florida USA


# On the Identification of Hysteretic Systems, Part II: Bayesian Sensitivity Analysis

**Keith Worden** and **Will Becker**

*Dynamics Research Group*
*Department of Mechanical Engineering*
*University of Sheffield*
*Mappin Street, Sheffield S1 3JD, UK*
*k.worden@sheffield.ac.uk*

**Abstract:** This paper forms the second in a short sequence considering the system identification problem for hysteretic systems. The basic model for parameter estimation is assumed to be the Bouc-Wen model, as this has proved particularly versatile in the past. Previous work on the Bouc-Wen system has shown that the system response is more sensitive to some parameters than others and that the errors in the associated parameter estimates vary as a consequence. The objective of the current paper is to demonstrate the use of a principled Bayesian approach to parameter sensitivity analysis for the Bouc-Wen system. The approach is based on Gaussian process emulation and is encoded in the software package Gem-SA. The paper considers a five-parameter Bouc-Wen model, and the sensitivity analysis is based on data generated by computer simulation of a single-degree-of-freedom system.

**Keywords:** Hysteresis, the Bouc-Wen model, system identification, uncertainty analysis, Bayesian sensitivity analysis.

## 1. Introduction

This paper represents the second in a short sequence discussing various aspects of the system identification problem for hysteretic systems i.e. systems with memory. As discussed in the first paper [1], an extremely versatile parametric form for the modeling of hysteretic systems is the Bouc-Wen model [2,3]. (As stated in [1], the literature on the Bouc-Wen model is extensive and the reader is referred to the recent book [4] for a comprehensive guide to the literature.) One of the main problems associated with adopting the Bouc-Wen form for a system model, is the identification of the model parameters; the problem is complicated considerably by nonlinearity and the presence of unmeasured states [5,6]. In the first paper in this sequence, it was shown that it is possible to frame the identification problem in terms of an optimisation problem which is amenable to solution by the use of evolutionary algorithms.

There are a number of problems associated with the identification problem. One difficulty is associated with the fact that the response of the given hysteretic system will generally be more sensitive to some parameters than others; and these parameters will, as a result, be estimated with lower accuracy. In a sense, this does not raise a problem. If the model is being developed solely as an effective representation of the system which will only be used to make predictions in similar operating conditions to those in which the identification data were obtained; then the insensitivity to the parameters means that any errors have little effect. However, if the purpose of the model is truly to infer the underlying physics in some sense, then the parameter values matter. More importantly, if a model is used to extrapolate, i.e. make predictions under conditions removed from those in which identification data were acquired, then the system may become sensitive to a different parameter set. If can also become important to estimate how uncertainty in the parameters propagate through to uncertainty in the response. Because of these issues, the problem of system identification becomes entwined with problems of *uncertainty analysis* (UA).

In its broadest terms, UA is concerned with quantifying output uncertainty given certain input uncertainties; this in turn requires some qualitative means of assigning a degree of uncertainty to a given situation. The most popular and longest-standing uncertainty theory is probability theory and this will provide the means for uncertainty analysis in this paper. A sub-problem of considerable interest within UA is *sensitivity analysis* (SA). This determines how individual input parameters are responsible for uncertainties in the output. Saltelli [7] divides SA approaches into three categories; in increasing order of power one has:

- The lowest-level analysis is *screening*, which simply ranks the inputs in order of their importance in affecting the output. This can help a model builder identify the set of most important inputs, as well as any inputs that contribute very little to the output (and can thus be eliminated from subsequent uncertainty analyses). The main drawback of screening is that it offers no quantification of effects beyond ranking.
- The next level, *local SA*, analyses and quantifies the effects of varying input parameters, but only around their immediate locality. This does not account for non-linear responses however, so is of limited use in complex models. The idea of local sensitivity analysis will be familiar to structural dynamicists in the context of finite-element model-updating [8]; in that case, it is possible to compute analytically, the derivative of an output with respect to specified inputs. The local nature of the process is obvious from the use of calculus.
- The most informative analysis is *global SA*, which investigates and quantifies uncertainties over the complete range of input space. Of course, this comes with the drawback of increased computational expense. In order to truly sample the complete range of possible parameters, one can always have recourse to Monte Carlo analysis [9], but this can be the most expensive option of all. If a single sample is acquired on the basis of a computationally expensive run of a large model, the accumulation of many samples for Monte Carlo analysis may not be feasible, certainly not if the dimension of the input set is high.

As discussed above, SA can be desirable for a number of reasons, most often to identify the most influential inputs whose uncertainty must be reduced in order to decrease output variance; however also for design optimisation, and model simplification – i.e. identifying parameters that have little or no effect on the output and can thus be discounted. A new approach which addresses the problem of the computational expense of global SA, is the use of Bayesian statistics. By creating an *emulator* or *surrogate model* or *metamodel* of the model under investigation, through fitting a Gaussian process to the response surface, sensitivity analysis data can be inferred at a greatly reduced computational cost, with little loss of accuracy. Computational savings can be up to one or two orders of magnitude [10]. The Bayesian approach is used in this paper to carry out a sensitivity analysis for the Bouc-Wen system. In fact, the outputs of that model are not expensive to acquire; however, there are other advantages to the Bayesian approach which make it an interesting one to consider.

The layout of the paper is as follows: Section Two briefly describes the Bouc-Wen model of hysteresis considered here and explains the evolutionary approach to system identification. Section Three discusses the theoretical basis of the Bayesian sensitivity analysis, including the background for the Gaussian process models which are used for the emulator. Section Four presents the results of the sensitivity analysis for the Bouc-Wen system and the paper concludes with a little discussion in Section Five.

## 2. Identification of the Bouc-Wen Hysteresis Model

As discussed in [1] the system of interest here will be the general Single-Degree-of-Freedom (SDOF) hysteretic system described in the terms of Wen [3], is represented below $z(y,\dot{y})$ the hysteretic part of the restoring force,

$$m\ddot{y} + c\dot{y} + ky + z(y,\dot{y}) = x(t) \tag{1}$$

where $m$, $c$ and $k$ are the mass, damping and stiffness parameters respectively, $y$ is the system response and $x(t)$ is the excitation; overdots denote differentiation with respect to time. The hysteretic component is defined by Wen [3] via the additional equation of motion,

$$\dot{z} = -\alpha\,|\,\dot{y}\,|\,z^n - \beta\dot{y}\,|\,z^n\,| + A\dot{y} \qquad \text{for } n \text{ odd}$$
$$\dot{z} = -\alpha\,|\,\dot{y}\,|\,z^{n-1}\,|\,z\,| - \beta\dot{y}z^n + A\dot{y} \qquad \text{for } n \text{ even} \tag{2}$$

The parameters $\alpha$, $\beta$ and $n$ govern the shape and smoothness of the hysteresis loop. The equations offer a simplification from the point of view of parameter estimation, in that the stiffness term in (1) can be combined with the $A\dot{y}$ term in the state equation for $z$. As a system identification problem, this set of equations presents a number of difficulties, foremost are: (a) the state $z$ is not measurable and therefore it is not possible to use equation (2) directly in a least-squares formulation, and (b) The parameter $n$ enters the state equation (2) in a nonlinear way; this means that a linear least-squares approach is not applicable to the estimation of the full parameter set.

Fortunately, there are a number of identification procedures which can accommodate the difficulties peculiar to the Bouc-Wen system; as discussed in [1], the differential evolution algorithm and it's adaptive variant offer a powerful approach to the parameter estimation. As in all evolutionary optimisation procedures, a population of possible solutions (here, the vector of parameter estimates), is iterated in such a way that succeeding generations of the population contain better solutions to the problem in accordance with the Darwinian principle of 'survival of the fittest'. The problem is framed as a minimisation problem with the cost function defined as a normalised mean-square error between the 'measured' data and that predicted using a given parameter estimate, i.e.

$$J(m,c,\alpha,\beta,A) = \frac{100}{N\sigma_y^2}\sum_{i=1}^{N}(y_i - \hat{y}_i(m,c,\alpha,\beta,A))^2 \tag{3}$$

where $\sigma_y^2$ is the variance of the 'measured' sequence of displacements $\{y_i, i = 1,\ldots,N\}$ and the caret denotes a predicted quantity. The question of interest is 'how sensitive is this cost function to the individual model parameters?' This question cannot be addressed using a classical approach to sensitivity analysis as it is not possible to give an explicit form for the cost as a function of the parameters which can then be differentiated. Although numerical differentiation could be used, such an approach would still be local. Both of these problems are overcome by the use of Bayesian sensitivity analysis.

Finally, some description of the data used here for illustration is given. The data for the current study were generated by numerical simulation. The coupled equations (1) and (2) were integrated forward in time in Matlab [12] using a fixed-step fourth-order Runge-Kutta scheme for initial value problems [13]. A better solution could potentially be found by using an adaptive solver like the $(4,5)^{\text{th}}$-order Runge-Kutta method for the solution encapsulated in the Matlab function *ode45* that can adjust the step-size in order to obtain a prescribed error tolerance. However, it was shown in [14], that the use of the adaptive scheme in the context of evolutionary system identification can lead to strange results. In reference [14], an attempt was made to picture the cost/fitness landscape for the identification problem. This involved trying to visualise the variations in the cost function as some subset of the parameters changed. If all but one parameter was held constant, the function that resulted gave some indication of the sensitivity to the given parameter. Figure 1 shows the value of the cost function at 201 points in the interval [-1.51,-1.49] for $\beta$ with all the other parameters held at their true values.

The result in Figure 1 is fascinating; the cost function is insensitive to $\beta$, essentially fluctuating about a constant value at this resolution, except for the drop to zero at the true value of $\beta$. This shows that the search for the minimum is likely to be very difficult. A further point of interest is the very fact that the data appears to be noisy. This is not expected to occur, as the cost function is deterministic. The 'problem' lies with the integration scheme; small variations in the point at which the solver meets the error criterion for a time-step, translate into small variations in the prediction, which become evident once the cost function has reached small values. It is clear that a classical sensitivity analysis based on numerical simulation could not work under these circumstances. In any case, as the optimisation scheme itself would have grave difficulty with the 'noise' on the cost function, a fixed-step Runge-Kutta scheme was adopted.

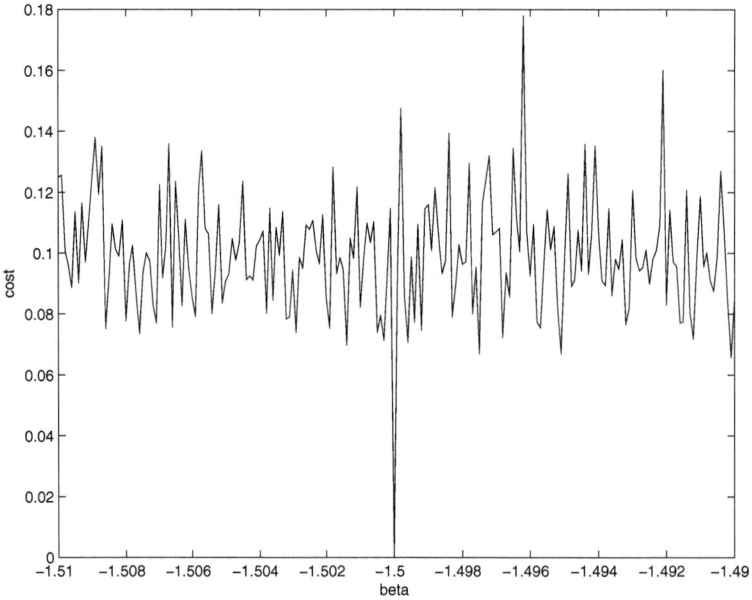


*Figure 1.* Cost as a function of $\beta$ with other parameters held at their true value.

The parameters for the baseline system adopted here were: $m = 1$, $c = 20$, $k = 0$, $\alpha = 1.5$, $\beta = -1.5$, $n = 2$ and $A = 6680$. The excitation $x(t)$ was a Gaussian random sequence with mean zero and standard deviation 9.92. These are the same parameter values as those adopted in [1] and [6]. The step-size (or sampling interval) was taken as 0.004 s, corresponding to a sampling frequency of 250 Hz. As the effect of noise was considered in [6], the current study uses uncorrupted data. The 'training set' or identification set used here was composed of 1000 points corresponding to a record duration of 4 seconds.

## 3. Bayesian Sensitivity Analysis

The SA technique here is based on a non-parametric probabilistic approach as detailed in [10]. Each uncertain input parameter is represented as a probability distribution, and a *Gaussian process* is fitted using multiple runs of the model as dictated by a Design-Of-Experiments (DOE). From this emulator, statistical quantities relating to sensitivity and uncertainty can be inferred directly – for example, output uncertainty distributions and main effects. Importantly, this requires no additional runs of the original model for each sensitivity measure, unlike conventional SA methods detailed for example in [7]. Furthermore, the advantage of using Gaussian process regression is that the uncertainty of the emulator fit is itself quantified, giving the analyst a very pragmatic quantification of the uncertainty in the data.

### *3.1. Gaussian Process Regression*

Any computer model can be regarded as a function of its inputs: $f(\mathbf{x})$. Although this function is deterministic and governed by known mathematical relationships, it is often of such complexity as to be considered mathematically intractable. So from a practical point of view, $f(\mathbf{x})$ could be regarded as an unknown function, given that one does not know the output for a given set of inputs until one has actually run the model. If, however, one samples the function (model) at a number of carefully chosen input points, it is possible to fit a response surface that can predict the output of the model for any point in input space without having to run the simulation itself. Although the idea of modelling a model (*metamodelling*) may seem a little abstract, for simulations that are computationally expensive it is a useful tool, since any approach for uncertainty analysis requires multiple runs of the model under investigation.

A particular approach to formulating the metamodel or emulator that has gathered interest in recent years is the use of Gaussian process regression [15-17]. Gaussian processes are an extension of a multivariate Gaussian probability distribution. Whereas most forms of regression return a crisp value $f(\mathbf{x})$ for any given $\mathbf{x}$, a Gaussian process returns a Gaussian probability distribution. Thus for a function, the Gaussian process can be considered as a multivariate Gaussian distribution, where the dimension of the multivariate distribution can be thought of as the 'resolution' of the function, or the number of predictive points. In the case detailed here, the resolution need not be specified, since one is not interested in predicting particular output points, but rather quantities pertaining to the whole range of output space.

Gaussian processes adhere to the Bayesian paradigm, that is, a number of prior assumptions are made about the function being modelled, and then training data (samples from the model) are used to update and evaluate a posterior distribution over functions. A key assumption is that the model is a smooth function of its inputs – it is this that allows extra information concerning the response to be gained at reduced computational cost.

For any set of $n$ input points $\{\mathbf{x}_1,\ldots,\mathbf{x}_n\}$ (which represent the values of the identification parameters for the specific problem considered here), each of dimension $d$, the prior beliefs about the corresponding outputs can be represented by a multivariate normal distribution, the mean of which is a least-squares regression fit through the training data,

$$E\{f(\mathbf{x}) \mid \beta\} = \mathbf{h}(\mathbf{x})^T \beta \tag{4}$$

where $\mathbf{h}(\mathbf{x})^T$ is a specified regression function of $\mathbf{x}$, and $\boldsymbol{\beta}$ is the corresponding vector of coefficients. For simplicity, $\mathbf{h}(\mathbf{x})^T$ was chosen here to be $(1, \mathbf{x}^T)$, representing a linear regression (this can be extended to higher polynomial fits if required). The covariance between output points is given as,

$$\operatorname{cov}\{f(\mathbf{x}), f(\mathbf{x}') \mid \sigma^2, B\} = \sigma^2 c(\mathbf{x}, \mathbf{x}') \tag{5}$$

where $\sigma^2$ is a scaling factor and $B$ is a diagonal matrix of length-scales, representing the roughness of the output with respect to the individual input parameters. The covariance function used here is chosen to be a squared-exponential function of the form,

$$c(\mathbf{x}, \mathbf{x}') = \exp\{-(\mathbf{x} - \mathbf{x}')^T B(\mathbf{x} - \mathbf{x}')\} \tag{6}$$

The posterior distribution is found by conditioning the prior distribution on the training data $\mathbf{y}$ (the vector of output points corresponding to the input set), and integrating out (or marginalising over) the hyperparameters $\sigma^2$ and $\beta$. The integrals involved are usually all Gaussian, and although the expressions are almost always very complicated, the results can be given in closed form. The result is a Student's t-process, conditional on $B$ and the training data,

$$[f(\mathbf{x}) \mid B, \mathbf{y}] \sim t_{n-q}\{m^*(\mathbf{x}), \hat{\sigma}^2 c^*(\mathbf{x}, \mathbf{x}')\} \tag{7}$$

where,

$$m^*(\mathbf{x}) = \mathbf{h}(\mathbf{x})^T \hat{\beta} + \mathbf{t}(\mathbf{x})^T A^{-1}(\mathbf{y} - H\hat{\beta}) \tag{8}$$

$$c^*(\mathbf{x}, \mathbf{x}') = c(\mathbf{x}, \mathbf{x}') - \mathbf{t}(\mathbf{x})^T A^{-1}\mathbf{t}(\mathbf{x}') + (\mathbf{h}(\mathbf{x})^T - \mathbf{t}(\mathbf{x})^T A^{-1}H)(H^T A^{-1}H)(\mathbf{h}(\mathbf{x}')^T - \mathbf{t}(\mathbf{x}')^T A^{-1}H)^T \tag{9}$$

$$\mathbf{t}(\mathbf{x})^T = (c(\mathbf{x}, \mathbf{x}_1), \ldots, c(\mathbf{x}, \mathbf{x}_n)),$$

$$H^T = (\mathbf{h}(\mathbf{x}_1), \ldots, \mathbf{h}(\mathbf{x}_n)),$$

$$A = \begin{pmatrix} 1 & c(\mathbf{x}_1,\mathbf{x}_2) & \dots & c(\mathbf{x}_1,\mathbf{x}_n) \\ c(\mathbf{x}_2,\mathbf{x}_1) & 1 & & \vdots \\ \vdots & & \ddots & \\ c(\mathbf{x}_n,\mathbf{x}_1) & \dots & & 1 \end{pmatrix}$$

$$\hat{\beta} = (H^T A^{-1} H)^{-1} H^T A^{-1}\mathbf{y}$$

$$\sigma^2 = \frac{\mathbf{y}^T\{A^{-1} - A^{-1}H(H^T A^{-1} H)^{-1} H^T A^{-1}\}\mathbf{y}}{(n-d-3)}$$

$$\mathbf{y}^T = (f(\mathbf{x}_1),...,f(\mathbf{x}_n))$$

Note that the determination of the emulator is basically an exercise in machine learning and therefore its quality is critically dependent on the number and distribution of training data points in the input space, and the values of the hyperparameters. The expressions for $\hat{\beta}$ and $\hat{\sigma}^2$ shown above represent least-squares estimates. The diagonal matrix of roughness parameters *B* cannot generally be integrated out analytically and is evaluated using maximum likelihood estimation – this calculation typically represents the most computationally intensive part of the process.

The dependence of the emulator on training data means that some model runs are always required. The advantage of the Bayesian sensitivity approach is that, typically far fewer runs are needed to train the emulator than would be needed for a full Monte Carlo analysis. To deal with the sampling of training data in as principled manner as possible, ideas of experimental design are applied and a *maximin Latin hypercube* design (maximin LHD) is used here. A Latin hypercube design divides input space into regions of equal probability and randomly assigns points that are distributed evenly across "probability-space". A maximin LHD improves on this by additionally maximising the minimum distance between input points, thus optimising the space-filling properties of the design.

*3.1. Inference for Sensitivity Analysis*

Several quantities can be inferred from the posterior distribution-over-functions described above, that are relevant to sensitivity analysis. Fundamental quantities such as the mean and variance of the output distribution can be evaluated, as well as *main effects*, *interactions* and sensitivity measures for input parameters based on their contribution to output variance. Additionally, the variance of the Gaussian process regression fit can be calculated – this represents the uncertainty about any of the results gained from the emulator.

Main Effects.The function $f(\mathbf{x})$ can be decomposed as follows, into *main effects* and *interactions*,

$$y = f(\mathbf{x}) = E(Y) + \sum_{i=1}^{d} z_i(x_i) + \sum_{i<j} z_{i,j}(x_{i,j}) + \sum_{i<j<k} z_{i,j,k}(x_{i,j,k}) + ... + z_{1,2,...,d}(\mathbf{x}), \quad (10)$$

where the expectation operator *E* is with respect to the probability distribution of the inputs and,

$$z_i(x_i) = E(Y \mid x_i) - E(Y)$$

$$z_{i,j}(\mathbf{x}_{i,j}) = E(Y \mid \mathbf{x}_{i,j}) - z_i(x_i) - z_j(x_j) - E(Y)$$

$$z_{i,j,k}(\mathbf{x}_{i,j,k}) = E(Y \mid \mathbf{x}_{i,j,k}) - z_{i,j}(\mathbf{x}_{i,j}) - z_{i,k}(\mathbf{x}_{i,k}) - z_{j,k}(\mathbf{x}_{j,k}) - z_i(x_i) - z_j(x_j) - z_k(x_k) - E(Y)$$

Here $z_i(x_i)$ represents the *main effect* of $x_i$, $z_{i,j}(x_{i,j})$ is the *first order interaction* and further terms represent higher order interactions; *Y* is the random variable corresponding to the function output *y* and therefore *E(Y)* is the expected value of the output $y$ considering all possible combinations of inputs. The main effect of an input can be thought of as the effect (on the output) of varying that parameter over its input range, averaged over all the other inputs. Interactions describe the effect of varying two or more parameters simultaneously, additional to the main effects of both variables. Plotting main effects serves as a visual indication of the influence of particular inputs and interactions, showing (albeit qualitatively) the variance of the output with respect to individual input parameters and the nonlinearities associated with those responses.

One can infer posterior mean values for main effects and interactions by simply substituting the posterior mean into the definitions detailed above. Since conditional expectation is defined as,

$$E(Y \mid \mathbf{x}_p) = \int_{\chi_{-p}} f(\mathbf{x}) G(\mathbf{x}_{-p} \mid \mathbf{x}_p) d\mathbf{x}_{-p} \tag{11}$$

where the subscripts $p$ and $-p$ denote the subset $p$ and the complement of $p$ respectively and $G(\mathbf{x})$ represents the multivariate probability distribution of the input parameters. The posterior mean of the main effect can be derived by substituting (8) into (11) in place of $f(\mathbf{x})$,

$$E^*\{E(Y \mid \mathbf{x}_p)\} = R_p(\mathbf{x}_p)\hat{\beta} + T_p(\mathbf{x}_p)\mathbf{e} \tag{12}$$

where $E^*$ is the expectation with respect to the posterior distribution now, and,

$$R_p(\mathbf{x}_p) = \int_{\chi_{-p}} \mathbf{h}(\mathbf{x})^T G_{-p|p}(\mathbf{x}_{-p} \mid \mathbf{x}_p) d\mathbf{x}_{-p} \tag{13}$$

$$T_p(\mathbf{x}_p) = \int_{\chi_{-p}} \mathbf{t}(\mathbf{x})^T G_{-p|p}(\mathbf{x}_{-p} \mid \mathbf{x}_p) d\mathbf{x}_{-p} \tag{14}$$

$$\mathbf{e} = A^{-1}(\mathbf{y} - H\hat{\beta})$$

Although this results in a series of matrix integrals, a Gaussian or uniform $G(\mathbf{x})$ distribution allows these to be solved analytically. Expressions for interactions can be similarly derived using their respective definitions. Note that when $p$ is the null set, (12) yields the posterior mean of the output i.e. $E^*\{E(Y)\}$.

Variance and Sensitivity Indices. Variance-based methods are widely used in sensitivity analysis. This involves quantifying the proportion of output variance for which individual input parameters are responsible. In particular, sensitivity can be measured by conditional variance:

$$V_i = \mathrm{var}\{E(Y \mid X_i)\}$$

This is the expected value of the contribution of the input variable $X_i$ to the output variance. Note that this is also the variance of the main effect of $x_i$, hence it is known as the *main effect index* (MEI). This can be extended to measure conditional variance of interactions of inputs, i.e. $V_{i,j} = \mathrm{var}\{z_{i,j}(\mathbf{x}_{i,j})\}$, and so on for higher order interactions. Although this approach allows detailed insight into the effects of combinations of inputs on output uncertainty, it can be time-consuming to examine all possible interaction permutations for models with many input dimensions. An alternative sensitivity measure [18], describes the output variance that would remain if one were to learn the true values of all inputs except $x_i$,

$$V_{Ti} = \mathrm{var}(Y) - \mathrm{var}\{E(Y \mid \mathbf{X}_{-i})\}$$

This measure, called the *total sensitivity index* (TSI), measures the variance caused by an input $x_i$ and any interaction of any order including $x_i$. It allows a more holistic view of the uncertainty attributed to each input, but does not give any details as to how it is distributed between main effects and interactions. Between the MEIs and TSIs, a detailed view of the sensitivities of inputs and their interactions can be gained.

To calculate posterior means and variances of the above quantities from the emulator, the following variance identity is used [10],

$$\mathrm{var}\{E(Y \mid \mathbf{X}_p)\} = E\{E(Y \mid \mathbf{X}_p)^2\} - E(Y)^2 .$$

Hence,

$$E^*(V_p) = E^*(E\{E(Y \mid \mathbf{X}_p)^2\}) - E^*(E(Y)^2) \tag{15}$$

$E^*(E(Y)^2)$ follows from the result,

$$E^*\{E(Y)^2\} = \mathrm{var}^*\{E(Y)\} + \{E^*(E(Y))\}^2 \tag{16}$$

This leaves $E^*[E\{E(Y \mid \mathbf{X}_p)\}]$, the equation for which is presented here:

$$\begin{aligned} &E^*[E\{E(Y \mid \mathbf{X}_p)\}] \\ &= \int_{\chi_p}\int_{\chi_{-p}}\int_{\chi_{-p}} E^*\{f(\mathbf{x})f(\mathbf{x}^*)\}G(\mathbf{x}_{-p} \mid \mathbf{x}_p)G(\mathbf{x}'_{-p} \mid \mathbf{x}_p)G(\mathbf{x}_p)d\mathbf{x}_{-p}d\mathbf{x}_{-p}d\mathbf{x} \\ &= \int_{\chi_p}\int_{\chi_{-p}}\int_{\chi_{-p}} E^*\{\hat{\sigma}^2 c^*(\mathbf{x},\mathbf{x}^*) + m^*(\mathbf{x})m^*(\mathbf{x}^*)\}G(\mathbf{x}_{-p} \mid \mathbf{x}_p)G(\mathbf{x}'_{-p} \mid \mathbf{x}_p)G(\mathbf{x}_p)d\mathbf{x}_{-p}d\mathbf{x}_{-p}d\mathbf{x} \end{aligned}$$

With the notation that **x*** is the vector comprising of $\mathbf{x}_p$ and $\mathbf{x}'_{-p}$. This can then be represented as the following,

$$\begin{aligned} &E^*[E\{E(Y \mid \mathbf{X}_p)\}] \\ &= \hat{\sigma}^2[U_p - \mathrm{tr}(A^{-1}P_p) + \mathrm{tr}\{W(Q_p - S_pA^{-1}H - H^TA^{-1}S_p^T + H^TA^{-1}P_pA^{-1}H)\}] \\ &+\mathrm{tr}(e^TP_p) + 2\mathrm{tr}(\hat{\beta}^TS_pe) + \mathrm{tr}(\hat{\beta}^TQ_p\hat{\beta}), \end{aligned} \tag{17}$$

where,

$$U_p = \int_{\chi_p}\int_{\chi_{-p}}\int_{\chi_{-p}} c(\mathbf{x},\mathbf{x}^*)G(\mathbf{x}_{-p} \mid \mathbf{x}_p)G(\mathbf{x}'_{-p} \mid \mathbf{x}_p)G(\mathbf{x}_p)d\mathbf{x}_{-p}d\mathbf{x}_{-p}d\mathbf{x},$$

$$P_p = \int_{\chi_p}\int_{\chi_{-p}}\int_{\chi_{-p}} t(\mathbf{x})t(\mathbf{x}^*)^TG(\mathbf{x}_{-p} \mid \mathbf{x}_p)G(\mathbf{x}'_{-p} \mid \mathbf{x}_p)dG(\mathbf{x}_p)d\mathbf{x}_{-p}d\mathbf{x}_{-p}d\mathbf{x},$$

$$Q_p = \int_{\chi_p}\int_{\chi_{-p}}\int_{\chi_{-p}} h(\mathbf{x})h(\mathbf{x}^*)^TG(\mathbf{x}_{-p} \mid \mathbf{x}_p)G(\mathbf{x}'_{-p} \mid \mathbf{x}_p)G(\mathbf{x}_p)d\mathbf{x}_{-p}d\mathbf{x}_{-p}d\mathbf{x},$$

$$S_p = \int_{\chi_p}\int_{\chi_{-p}}\int_{\chi_{-p}} h(\mathbf{x})t(\mathbf{x}^*)^TG(\mathbf{x}_{-p} \mid \mathbf{x}_p)G(\mathbf{x}'_{-p} \mid \mathbf{x}_p)G(\mathbf{x}_p)d\mathbf{x}_{-p}d\mathbf{x}_{-p}d\mathbf{x}.$$

The full details of the calculation of $V_i$ and $V_{Ti}$ can be found in [10]. All the quantities of interest presented here were computed using the software package Gem-SA [19].

## 4. Results of Sensitivity Analysis for Identification of Bouc-Wen System

The object of the exercise here is to infer how sensitive the model error (cost function) in the identification problem is to the individual model parameters; this will ultimately yield information about the likely errors in the parameter estimates. In order to estimate the sensitivity measures described in the previous section, it is necessary to specify the probability distributions for the input parameters. As the differential evolution strategy assumes initial bounds on the parameters, it makes sense to specify a uniform distribution for the parameters within these bounds. This is fortunate, as the Gem-SA software accommodates uniform distributions. In the first stage of the identification scheme, one usually has little prior knowledge of the parameters and this will be reflected here by assuming bounds on the parameters which range over an order of magnitude above and below the true values; this means the ranges are: [0.1,10.0] for $m$, [2.0,20.0] for $c$, [0.15,15.0] for $\alpha$, [-15.0,-0.15] for $\beta$ and [668.0,66800.0] for $A$. In order to apply the emulator-based approach, it was necessary to specify training data; 400 combinations of input parameters were generated within Gem-SA using a Latin hypercube design and the corresponding hysteretic systems responses and cost functions were computed using Matlab. Of the 400 runs, three of the parameter combinations gave divergent response calculations and these data points were removed from the training set. The sensitivity analysis was then carried-out. In order to test the fidelity of the Gaussian process emulator at the heart of the analysis, 20% of the data were reserved as a cross-validation test set.

The sensitivity analysis here corresponds to large uncertainty on the input set. As the vast majority of the input parameters here correspond to poor identification results, one would expect the cost function to be high over most of the input space; this is reflected in the mean of the cost function computed in the sensitivity analysis, one finds that $E(Y) = 95.66\%$ for this particular run. Before considering the sensitivity results, it is wise to look at the accuracy of the emulator model; this is reflected in Figure 2 which shows the emulator predictions (and their corresponding confidence bands) on the test set (data not used in training).

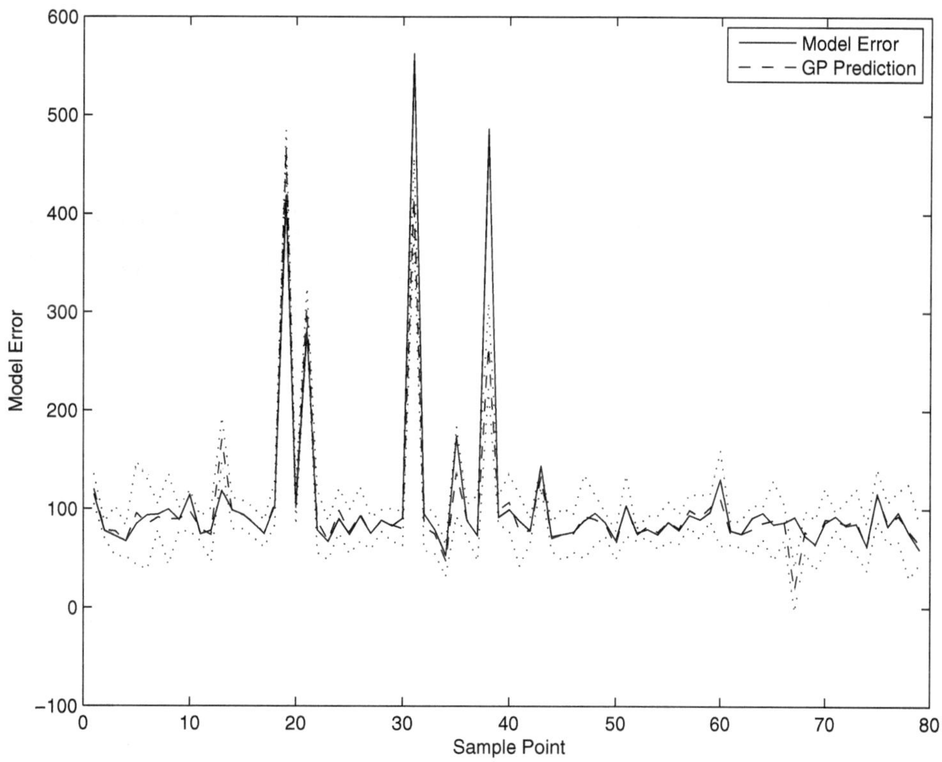


*Figure 2.* Emulator model predictions compared to true values of cost function on cross-validation test set for broad ranges of parameters (95% confidence bounds on predictions are shown by dotted lines).

The results in Figure 2 indicate that the emulator is doing a good job; in quantitative terms the RMS error over the cross-validation set is 32.7361 and this is largely due to large errors on the prediction of very high values of the cost function. The main effects plots for the analysis are given in Figure 3. These show the effect of the individual parameter uncertainties in driving the output uncertainty. In each case, multiple curves are shown, corresponding to random samples of the other parameters and this gives an idea of the effect of interactions between parameters.

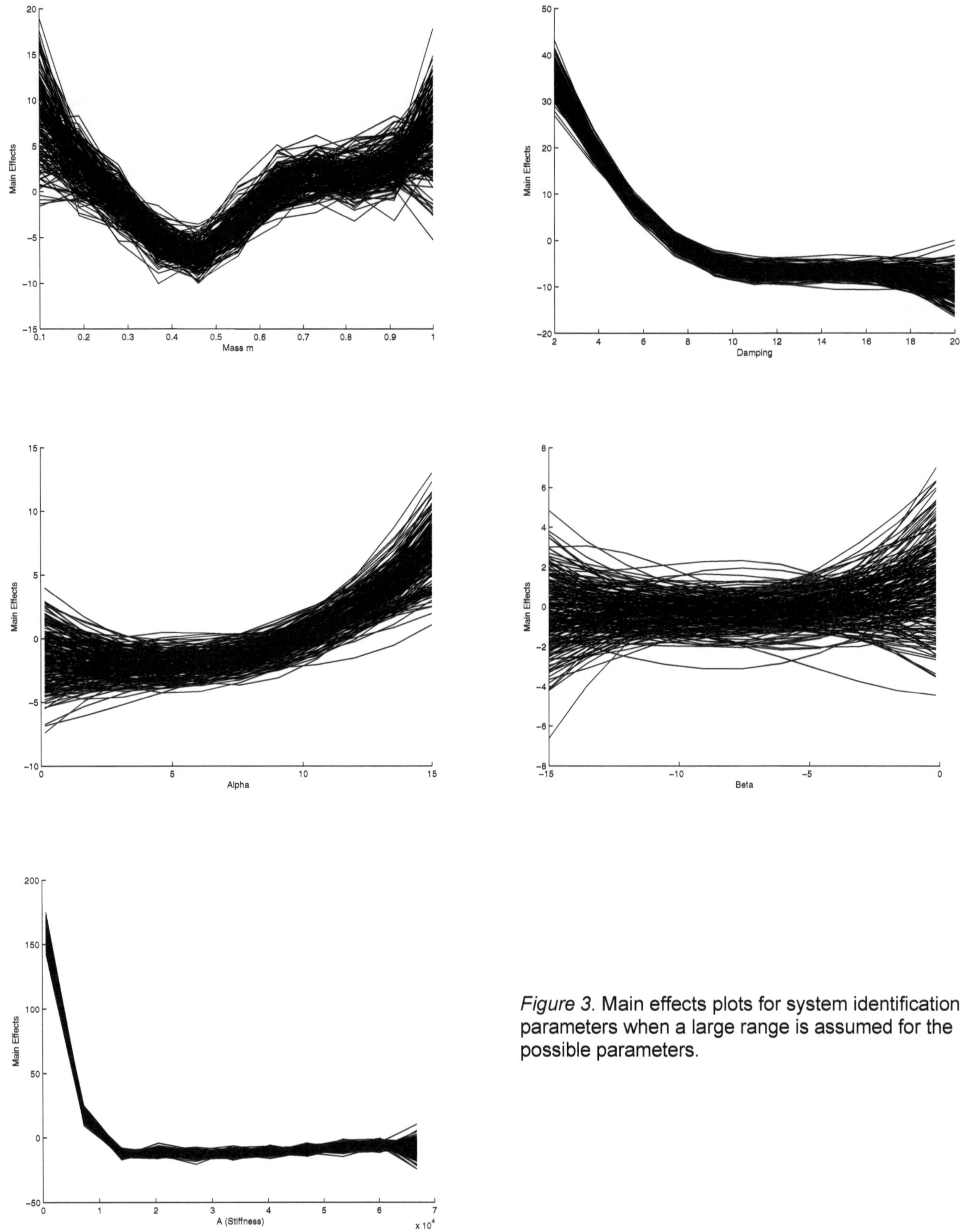


*Figure 3.* Main effects plots for system identification parameters when a large range is assumed for the possible parameters.

The main effects should be considered in the contest of the mean output, which is in this case 95.7. Figure 3 allows a number of interesting observations. First of all, the curves are not 'constants', this indicates nonlinearity or the fact that sensitivity to a given parameter depends on the parameter itself. This is most marked in the case of the $A$ parameter which shows a very high effect on the output at low values – corresponding to higher flexibilities of the system, a slightly milder (but still very significant) nonlinearity is clear for the damping parameter $c$. A further interesting observation is that the main effects for the Bouc-Wen parameters corresponding to the nonlinear terms - $\alpha, \beta$ - are comparatively low, suggesting that the linear dynamics of the system is dominant. The results can also be summarised as in Tables 1 and 2.

| Input Parameter | Variance (%) | Total Effect |
|---|---|---|
| $m$ | 0.81 | 31.28 |
| $c$ | 5.59 | 36.40 |
| $\alpha$ | 0.32 | 9.56 |
| $\beta$ | 0.04 | 4.54 |
| $A$ | 40.84 | 90.92 |

*Table 1.* Sensitivity analysis for identification problem with large bounds on parameters: main effects.

| Input Parameters | Variance (%) | Input Parameters | Variance (%) |
|---|---|---|---|
| $m, c$ | 0.73 | $c, \beta$ | 0.06 |
| $m, \alpha$ | 0.54 | $c, A$ | 18.42 |
| $m, \beta$ | 0.18 | $\alpha, \beta$ | 0.05 |
| $m, A$ | 14.13 | $\alpha, A$ | 1.45 |
| $c, \alpha$ | 0.11 | $\beta, A$ | 0.30 |

*Table 2.* Sensitivity analysis for identification problem with large bounds on parameters: interactions.

Table 1 summarises the main effects for the parameters, it is understood as follows. Uncertainty in the parameters will generate uncertainty in the output, in this case the cost function; the 'variance' column gives the contribution to the output variance from a given parameter; a high value here means that the output variance will be reduced considerably if one were to learn the true value of the given parameter. (Note that the variances will not generally sum to 100% because of interactions.) It appears here that the uncertainty in the stiffness - $A$ - and damping - $c$ - parameters is driving the output uncertainty; the nonlinear Bouc-Wen parameters appear to make a minor contribution, particularly $\beta$; this supports the information from Figure 1. As indicated above, the main effects are not the whole story; the column 'total effect' represents the contribution from a parameter if one includes all its interactions with other parameters. When interactions are taken into account, the mass parameter begins to appear significant. This is entirely to be expected as the mass parameter in combination with the stiffness and damping parameters fixes the natural frequency and damping ratio of the system; the interaction effects are shown clearly in Table 2.

A second sensitivity analysis was performed for a narrower range of parameters; this was in order to consider the situation where an earlier identification run might have produced better bounds on the parameters. For the second analysis, the ranges were taken as: [0.5,1.5] for $m$, [10.0,30.0] for $c$, [0.75,2.25] for $\alpha$, [-2.25,-0.75] for $\beta$ and [3340.0,10020.0] for $A$. As before, a uniform distribution was assumed for each parameter over these ranges and training data for the emulator were generated using a Latin hypercube design. In this case, one finds $E(Y) = 70.07\%$, indicating that lower cost functions values are obtained for the narrower intervals around the true values; this is entirely to be expected. As before, it was considered sensible to consider the accuracy of the emulator fit; the results of the predictions on a cross-validation set are shown in Figure 4.

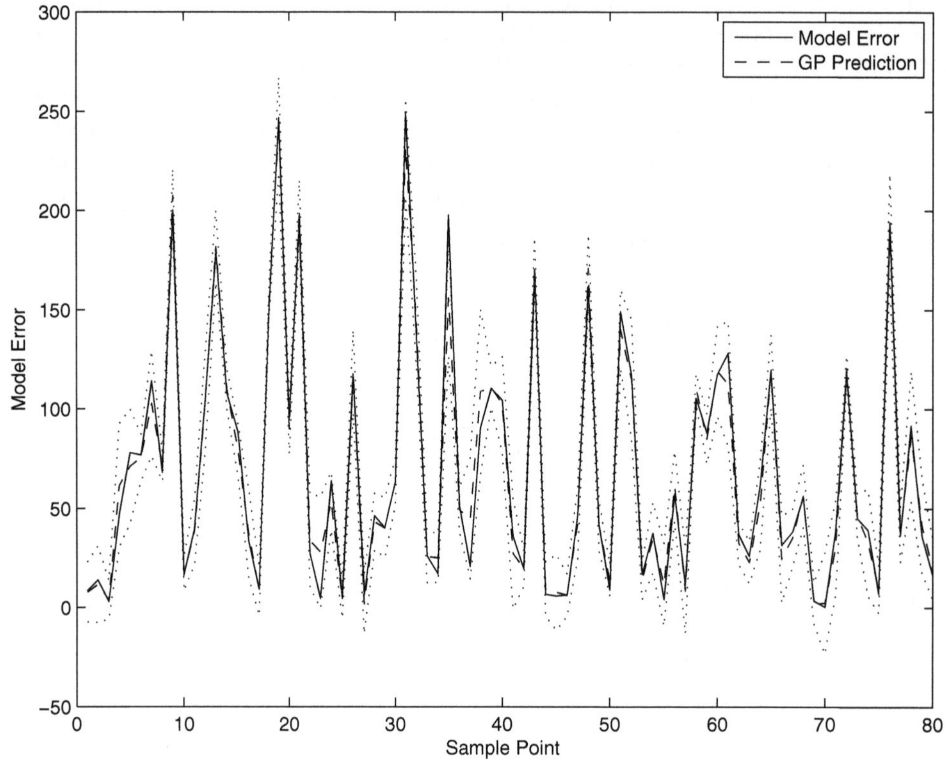


*Figure 4.* Emulator model predictions compared to true values of cost function on cross-validation test set for broad ranges of parameters (95% confidence bounds on predictions are shown by dotted lines).

It is clear from Figure 4 that the emulator has captured very well the behaviour of the cost function. The summaries for the main effects and interactions can be seen in Figure 5 and Tables 3 and 4.

| Input Parameter | Variance (%) | Total Effect |
|---|---|---|
| $m$ | 7.73 | 66.19 |
| $c$ | 14.63 | 20.29 |
| $\alpha$ | 0.04 | 1,51 |
| $\beta$ | 0.03 | 1.46 |
| $A$ | 17.63 | 76.41 |

*Table 3.* Sensitivity analysis for identification problem with narrow bounds on parameters: main effects.

| Input Parameters | Variance (%) | Input Parameters | Variance (%) |
|---|---|---|---|
| $m,c$ | 0.46 | $c,\beta$ | 0.03 |
| $m,\alpha$ | 0.06 | $c,A$ | 0.89 |
| $m,\beta$ | 0.07 | $\alpha,\beta$ | 0.03 |
| $m,A$ | 53.29 | $\alpha,A$ | 0.05 |
| $c,\alpha$ | 0.05 | $\beta,A$ | 0.09 |

*Table 4.* Sensitivity analysis for identification problem with narrow bounds on parameters: interactions.

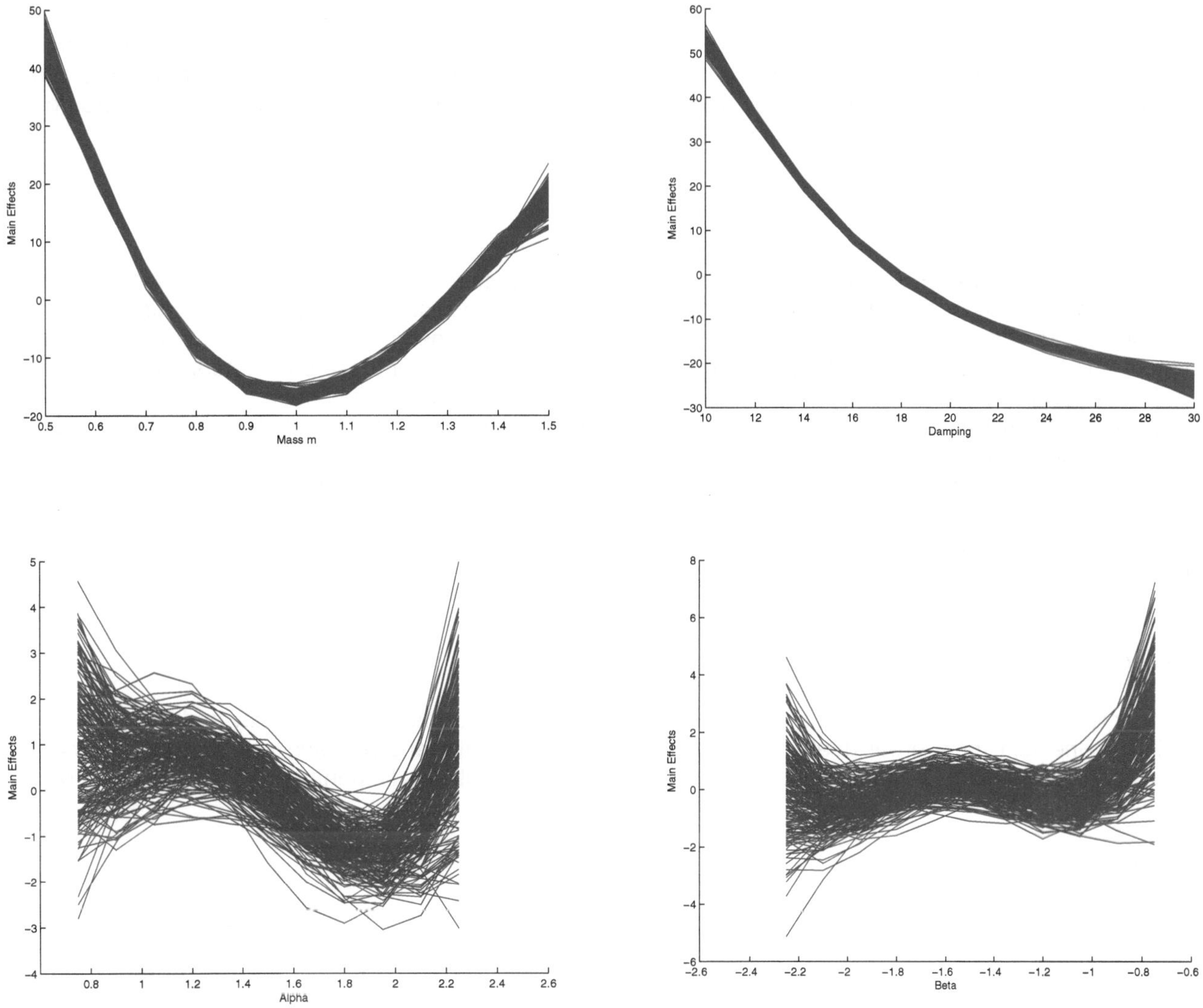


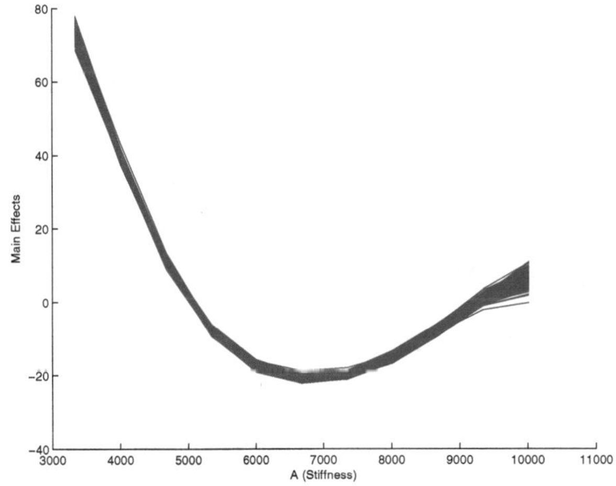


*Figure 5.* Main effects plots for system identification parameters when a narrow range is assumed for the possible parameters.

The results are broadly similar to those for the larger range of parameters. Again there is marked nonlinearity for the stiffness and damping main effects; in this case the main effect for the mass also shows severe nonlinearity. The only real differences in terms of the main effects etc. here is that the uncertainty in the damping parameter assumes a lesser role if the parameter bounds are narrower i.e. severe damping errors generate large output errors, but mild damping errors are much less significant. As before, the nonlinear Bouc-Wen parameters do not appear to be substantial drivers of the output uncertainty.

## 5. Conclusions

The conclusions of the paper are rather straightforward. The Bayesian sensitivity analysis has been shown to shed light on the system identification problem for Bouc-Wen hysteretic systems. Perhaps a little surprisingly, the analysis seems to indicate that the identification problem is rather insensitive to the nonlinear Bouc-Wen parameters and this provides support for the observation made elsewhere that these parameters appear to be the most difficult to estimate with accuracy. One of the main objectives of this paper has been to demonstrate the use of the Bayesian sensitivity analysis in a structural dynamic context as the authors believe that it provides a principled framework for the analysis of uncertainty in many important dynamical systems problems. With this in mind, the Bouc-Wen system has been chosen as an illustration because of its relative simplicity; however, it is intended to apply the method to recently developed extensions of the Bouc-Wen model which include many more parameters. The next paper in this short sequence will be concerned with the estimation of confidence intervals for the parameter estimates obtained from the evolutionary identification scheme.

## Acknowledgements

The authors would like to sincerely thank Dr Jeremy Oakley of the Department of Probability and Statistics at the University of Sheffield for a great deal of help and advice on sensitive analysis and also on general matters Bayesian; thanks also go to Dr Marc Kennedy of the Probability and Statistics Department for the use of GEM-SA.

## References

[1] Worden, K., and Manson, G., "On the Identification of Hysteretic Systems, Part I: An Extended Evolutionary Scheme", *Proceedings of xth IMAC Conference, Jacksonville, Florida, US*, 2010.

[2] Bouc, R., "Forced Vibration of Mechanical System With Hysteresis", *Proceedings of 4th Conference on Nonlinear Oscillation, Prague, Czechoslovakia*, 1967.

[3] Wen, Y., "Method for Random Vibration of Hysteretic Systems", *American Society of Civil Engineers - Journal of the Engineering Mechanics Division,* XX, 1976, pp.249-263.

[4] Ikhouane, F., and Rodellar, J., *Systems with Hysteresis: Analysis, Identification and Control using the Bouc-Wen Model*, Wiley-Blackwell, 2007.

[6] Kyprianou, A., Worden, K. and Panet, M., "Identification of Hysteretic Systems using the Differential Evolution Algorithm", *Journal of Sound and Vibration*, **248**, 2001, pp.289-314.

[7] Saltelli, A.K., Chan, E.M. Scott, *Sensitivity analysis,* Wiley, New York, 2000.

[8] Friswell, M.I. and Mottershead, J.E., *Finite Element Model Updating for Structural Dynamics*, Kluwer Academic Publishers, 1995.

[9] Shreider, Y., A., *Method of statistical testing: Monte Carlo method,* Elsevier, Amsterdam ; London (1964).

[10] Oakley, J.E. and O'Hagan, A., "Probabilistic Sensitivity Analysis of Complex Models: a Bayesian Approach", *Journal Of The Royal Statistical Society Series B*, **66**, 2004. pp.751-769.

[11] Price, K. and Storn, R., "Differential Evolution – a Simple and Efficient Heuristic for Global Optimization over Continuous Spaces", *Journal of Global Optimization*, **11**, 1997, pp.341-359.

[12] Matlab v7, The Mathworks, 2004.

[13] Press, W.H., Teukolsky, S.A., Vetterling, W.T. and Flannery, B.P., *Numerical Recipes 3rd Edition: The Art of Scientific Computing*, Cambridge University Press, 2007.

[14] Worden, K., "The Fitness Landscape for Parameter Estimates for Hysteretic Systems", *in Proceedings of 2nd International Conference on Engineering Dynamics – ICEDyn09, Ericiera, Portugal*, 2009.

[15] Kennedy, M.C., Anderson, C.W., Conti, S. and O'Hagan, A., "Case Studies in Gaussian Process Modelling of Computer Codes", *Reliability Engineering and System Safety*, **91**, 2006, pp.1301-1309.

[16] Kennedy, M.C. and O'Hagan, A., "Bayesian Calibration of Computer Models", *Journal of the Royal Statistical Society: Series B*, **63**, 2001, pp.425-464.

[17] Sacks, J., Welch, W.J., Mitchell, T.J. and Wynn, H.P., "Design and Analysis of Computer Experiments", *Statistical Science*, **4**, 1989, pp.409-435.

[18] Homma, T. and Saltelli, A.K., "Importance measures in global sensitivity analysis of model outputs", *Reliability Engineering and System Safety*, **52**, 1996. pp.1-17.

[19] Kennedy, M., *Gem-SA Homepage*. [Available from: http://ctcd.group.shef.ac.uk/gem.html].

Proceedings of the IMAC-XXVIII
February 1–4, 2010, Jacksonville, Florida USA


# Identifying and Quantifying Structural Nonlinearities from Measured Frequency Response Functions

A. Carrella[1], D.J. Ewins[1], A. Colombo[2], E. Bianchi[2]

[1]University of Bristol, BLADE / Department of Aerospace Engineering, Bristol, UK

[2]AgustaWestland, Cascina Costa, Italy

**Abstract**

Engineering structures seldom behave linearly and linearity checks are common practice in the development of critical structures exposed to dynamic loading to define the boundary of validity of the linear regime. However, and especially in large scale industrial applications, there is not a general methodology for dynamicists to extract nonlinear parameters from measured vibration data so that these can be then included in the associated numerical models. In this paper, it is proposed a simple method based on the information contained in the frequency response function (FRF) of a structure. The technique falls within the category of Single-Degree-of-Freedom (SDOF) methods. The basic principle is that, at a fixed amplitude of response, it is possible to extract the dynamic properties of the underlying linear system. Repeating this linearisation for different response amplitudes allows to extract the stiffness and damping as functions of the amplitude of vibration. Furthermore, because of its mathematical simplicity and practical implementation during standard vibration test, is particularly suitable for the engineering community. Results of numerical simulations as well as measured data demonstrate the potential benefit offered by this approach.

# 1 Introduction

In order to asses and to predict the behaviour of a structure, engineers rely ever more heavily on mathematical models of the system under examination. These mathematical models can be analytical (e.g. a set of equations) or numerical - e.g. a Finite Element (FE) mesh. Either way, a model has to reproduce the reality. Therefore, notwithstanding the dramatic improvements in simulation techniques, testing is always required to validate predictions and to guide the refinement of models used for design optimisation by a procedure known as Model Updating.

One major drawback of model updating (and modal analysis) is that in most cases neither the mathematical model nor the analysis of measured data account for nonlinearities into the system. Mainly, this is due to the combination of a relatively small understanding of nonlinear phenomena, together with a highly complex mathematical analysis.

In reality it is found that most practical structures do not comply with the assumption of linearity. There are available a series of 'linearity checks' that can be performed in order to assess and define the limits of the validity of linear theory. The questions is what happened if the structure fails the linearity check. There is not a definite answer, and three different strategies can be envisaged: 1) the nonlinearity is completely ignored and the modal analysis is effected with standard linear methods - specifying the limits of validity of the modal parameters; 2) a full-scale nonlinear investigation can be initiated. Although accurate and exact, this requires a set of very time-consuming and mathematically complex tools which are more suitable to researchers in academia than engineers in industrial community; 3) finally, the nonlinearity can be detected, acknowledged and, even if at a first degree of approximation, the nonlinear parameters can be extracted and used to improve the quality of the numerical or mathematical model.

Consider, for example, the nonlinear system with quadratic (hardening) stiffness

$$m\ddot{x} + c\dot{x} + k\,x + k_{nl}\,x^3 = F_0\,\sin(2\pi\,f_e\,t) \tag{1}$$

with $m$ =1.5 kg, $c = 0.8$ Ns/m, $k = 6000$ N/m, $k_{nl} = 7\text{x}10^6$ N/m$^3$ and $F_0 = 0.2$N. By varying the frequency of excitation $f_e$ and solving numerically Eqn.(1) the FRF of the system can be computed. Fig.1 shows the results of the circle-fit method - a standard linear modal analysis technique.

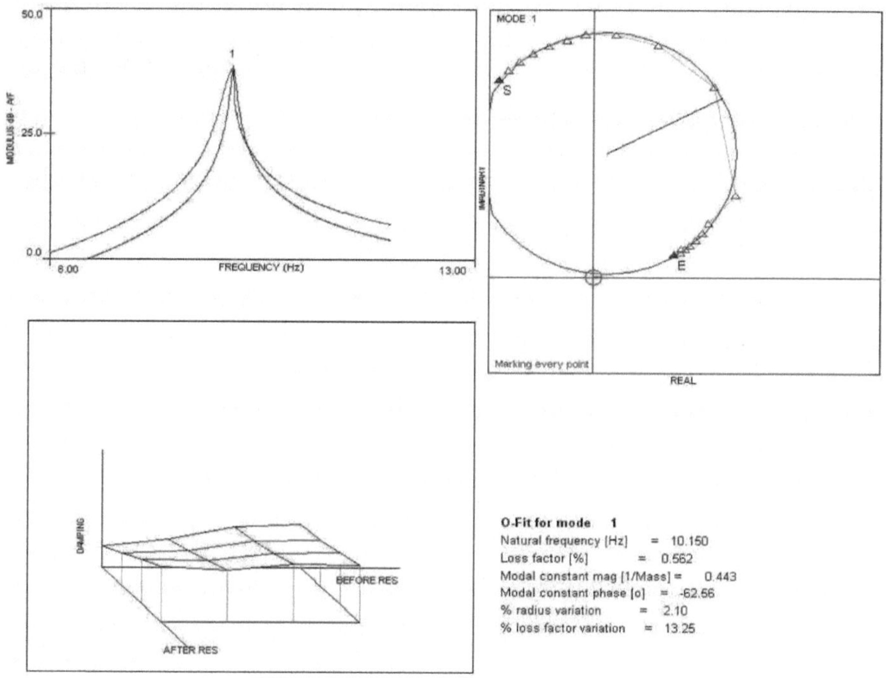


Fig. 1: Modal Analysis of a the nonlinear system with cubic hardening nonlinearity described by Eqn.(1). The methods used is the standard circle fit method as implemented by a commercially available software

There are three clear signs that applying a linear modal analysis method to a nonlinear system yields erroneous results: a) the curve fit of the FRF give a poor match (top-left figure); b) the damping carpet is distorted even is the damping model used is linear; c) there are -62.56° of complexity in the mode considered (the cause of which can only be imputed to the nonlinear term). At the same time it should be noted that the approach 2) - a full-scale nonlinear analysis - is also not applicable to the needs of the majority of the engineers because of the time and instruments required for a comprehensive mathematical analysis. It is clear there is scope for proposing a methodology that

- detect nonlinear behaviour form standard linear modal tests;
- analyse measured data so as to obtain an estimate of the type and magnitude of nonlinearity(ies);

Establishing the current state-of-the-art in the field of identification and quantification of nonlinearities in structural dynamics, herein also referred to as NonLinear Modal Testing (NLMT) is quite a demanding task. Fortunately, there exist some recent references which present a comprehensive and detailed collection of the different methods and techniques developed in the past three decades or so. The reference textbook, if not the only one, for nonlinear modal analysis was published in 2001 and written by Worden and Tomlinsen [1]. Some years later Kershen *et al* [2] published a review paper in which 446 references were studied. As also the authors stated, their review has inevitably missed some works on the subject. Nonetheless, reference [2] is an authoritative paper which set the basis for any study on nonlinear dynamic testing thereafter. It also important to note that, for obvious reasons of commercial interest, there is scarce number of works published on NLMT that refers to industrial research and/or practice. A reference which addresses the more general engineering community with the aim of "standardising laboratory practice in dynamic testing" is the nine-volume Dynamic Testing Agency (DTA) Handbook. In particular, volume 4 refers to Nonlinearity in Dynamic Testing [3] - which interestingly discuss most of the methods presented in [1]. Because the Handbook is aimed at industrial practice, in the tome all aspects (from set-up to signal processing) of performing a dynamic test aimed at detecting, quantifying and characterising nonlinearities are described. However, the work is more than 10 years old and therefore does not account for the most recent developments.

In view of the comprehensive review paper [2] and the reference textbook [1], it is might be redundant to present an overview of the different techniques used for NLMT. It should be noticed that, as also pointed out in ref. [1], a universal method for NLMT does not exist (and is unlikely to be developed in the near future): every method presently available has its own strength and weaknesses and, depending on the application, some can be more suitable than another. Each method presently available has its own strengths and weaknesses and, depending on the application, some can be more suitable than others. In the authors's view, amongst those available in the literature, the 3 methods most suitable to the engineering community for their trade-off between mathematical complexity and practical applicability are the **Restoring Force Surface (RFS)**, [1, 4], the **Inverse Method** [1, 5, 6] and the **Linearity Plots**, [7]. A critical review of the first two methods can be found in the cited references, however, the linearity plot deserves a special mention. In fact, references [8–12] show that (i) nonlinearities cannot be neglected during the dynamics analysis of large structures, and (ii) that linearity plots are the current tool for nonlinear analysis used in industry. The major drawback is the lengthy process involved in measuring the number of FRFs (at different level of excitation) required for producing the linearity plots.

It can be concluded that there is need and scope to develop a simple tool which allows to extract structural nonlinearities from standard vibration measurement, at least to a first degree of approximation. Such a technique would be of significant benefit to the modelers which could then accommodate these nonlinearities in their predictive analysis.

## 2 Nonlinear identification and characterisation

In reference [13] Lin outlines a method for identifying the presence of structural nonlinearities from measured FRFs. Its strengths are the simple mathematics and its implementation and applicability to standard vibration measurements. The Single-Degree-Of-Freedom (SDOF) method for identification and characterisation of structural nonlinearities described in this paper, extends Lin's observations (identification) to include information extracted from the properties of the inverse receptance to infer the spatial properties (characterisation). In order to verify the approach, the numerical FRFs of a SODF system with typical nonlinearities will be generated. The coefficient for the numerical examples are chosen to be equal to those in reference [7]. It is shown that the parameters extracted provide a good approximation of the system investigated. Finally, the method will be applied to experimental data.

### 2.1 Nonlinear identification

The dynamics of a nonlinear system with amplitude-dependant damping and/or stiffness - which are the most common classes of nonlinearities in engineering structures - excited by a harmonic force is expressed by the solution of its equation of motion

$$m\ddot{x} + c(X)\dot{x} + k(X)\,x = \Gamma_0\,\sin(2\,\pi\,f_e\,t) \tag{2}$$

where $X$ is the amplitude of the response. Note that a linear system is a special case of Eqn.(2) in which the functions $c(X)$ and $k(X)$ are constants. Firstly, it should ne mentioned that, for this type of nonlinearities, the influence on the measured data can be minimised by conducting tests in such a way that the response amplitude is kept constant through any frequency region where a non-linearity is likely to be activated. In these level-controlled sinusoidal tests the excitation level is adjusted such that the response amplitude is maintained at a chosen controlled value and have the effect of linearising the behaviour of the structure and of yielding the properties of the underlying linear model of the structure which is valid for that particular amplitude of vibration. Repeating the tests at different amplitudes provide a measure of the non-linear characteristic of the structure. However, the implementation of this technique is a rather complex matter because of the difficulties encountered in controlling the force level that produces a constant amplitude of oscillations [6].

The proposed identification method is based on a different linearisation approach. Without imposing any control or restriction on the excitation (although it would be preferable to maintain a constant force level throughout the test), at a given response amplitude $X$, the functions $c(X)$ and $k(X)$ in Eqn.(2) are in effect constants. In other words, it is possible to linearise the system at that specific amplitude. The implication of this linearisation is 2-fold:

1. it is licit to assume that the response of the system *at that amplitude* has the same frequency of the excitation, i.e.

$$x(t) = X \sin(2\pi f_e t + \varphi) \tag{3}$$

2. at any given amplitude it is possible to define the receptance as [6]

$$H(\omega) = \frac{X(\omega)}{F(\omega)} = \frac{A_r + \jmath B_r}{\omega_r^2(X) + \omega^2 + \jmath \eta_r(X)\omega^2} \tag{4}$$

in which $A_r$ and $B_r$ are the real and imaginary parts of the modal constant respectively, $\omega_r(X)$ and $\eta_r(X)$ are the natural frequency and the modal loss factor at that given amplitude. The identification process consists in extracting the functions of $\omega_r(X)$ and $\eta_r(X)$. This can be done in the following manner by measuring any two of the three quantities $H(\omega)$ (which is complex for any damped structure), $X(\omega)$ and $F(\omega)$.

Consider the plot of the displacement amplitude against frequency in Fig.2.

This has been obtained by multiplying the receptance by the force spectrum, $X(\omega) = H \times F$. At any given amplitude, $X_i$ say, there is a complex pair of points (1 and 2 in the figure) that defines the FRF. The amplitude of the receptance $|H|$ is related to the real and imaginary parts

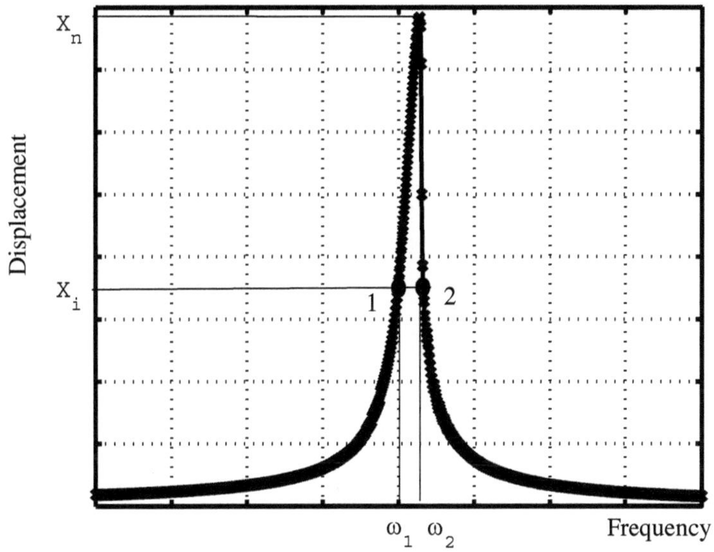


Fig. 2: Typical nonlinear FRF. At a given amplitude $X_i$ the is a pair of complex point on the FRF which contains the information required to calculate the natural frequency and the damping at that particular amplitude. By repeating the calculation at different response amplitudes $X_i$ $i$ = 1..n, it is possible to construct the functions $\omega_r(X)$ and $\eta_r(X)$ according to Eqns.(6)

of the 2 FRF points by [6]

$$H = \frac{(A_r + \jmath\, B_r)}{\omega_r^2 + \omega_{1i}^2 + \jmath\, \eta_r \omega_{1i}^2} = R_{1i} + \jmath\, I_{1i} \tag{5a}$$

$$H = \frac{A_r + \jmath\, B_r}{\omega_r^2 + \omega_{2i}^2 + \jmath\, \eta_r \omega_{2i}^2} = R_{2i} + \jmath\, I_{2i} \tag{5b}$$

Combining Eqns.(5a) and (5b) it is possible to construct the functions $\omega_r(X)$ and $\eta_r(X)$ as follow[1]

$$\omega_r^2(X) = \frac{(R_2 - R_1)\,(R_2\,\omega_2^2 - R_1\,\omega_1^2) + (I_2 - I_1)\,(I_2\,\omega_2^2 - I_1\,\omega_1^2)}{(R_2 - R_1)^2 + (I_2 - I_1)^2} \tag{6a}$$

$$\eta_r(X) = \left| -\frac{(I_2 - I_1)\,(R_2\,\omega_2^2 - R_1\,\omega_1^2) + (R_2 - R_1)\,(I_2\,\omega_2^2 - I_1\,\omega_1^2)}{\omega_r^2\,(R_2 - R_1)^2 + (I_2 - I_1)^2} \right| \tag{6b}$$

Finally, computing Eqn.(6) at different amplitude values ($X_i$, $i$ =1..n) allows to define the variation of the natural frequency and loss factor with the amplitude of vibration.

### 2.2 Nonlinear characterisation

The identification process outlined above, allows the identification of a potential nonlinearity. The next step is to quantify the nonlinear function. For this purpose there is need of converting

[1]The subscript $_i$ is omitted for sake of clarity

| Nonlinearity | Damping force | Restoring force | Values |
|---|---|---|---|
| Hardening Cubic Stiffness | $c\,\dot{x}$ | $k\,x + k_{nl}\,x^3$ | $k = 6000$ N/m<br>$k_{nl} = 7\text{e}^6$ N/m$^3$ |
| Quadratic Damping | $c\,\dot{x} + c_{nl}\,\dot{x}\,\lvert\dot{x}\rvert$ | $k\,x$ | $k = 6000$ N/m<br>$c_{nl} = 8Ns/m$ |

Table 1: Description of the nonlinearity and values of the coefficients

the modal quantities, or functions (natural frequency and loss factor), into spatial functions $k(X)$ and $c(X)$ so that the SDOF model can be put in the form of Eqn.(2). Firstly, there is need of computing the (modal) mass of the system, $m$. This can be calculated from the slope of the plot of the real part the inverse of the receptance against the square of the frequency [6]. The information on the mass of the system, together with the natural frequency (now a function of displacement, as shown above) provides the stiffness function $k(X)$. Also, the damping function can be extracted using the relationship $c(X) = \eta(X)\,m\,\omega_r(X)$. Note that in so doing it has been made the approximation that $\eta \approx 2\zeta$ which is strictly true only for a linear system at resonance.

## 3 Numerical simulation

In order to show the validity of the approach the method is first applied to 2 numerical examples. Two different nonlinearities - amongst those typically encountered in engineering applications - are a hardening cubic stiffness and a quadratic damping. The equation of motion of the SDOF system excited by a harmonic force is written as

$$m\ddot{x} + f_c\dot{x} + f_k\,x = F_0\,\cos(\omega_e\,t) \tag{7}$$

where the numerical values are taken from ref.[7]: the mass of the system is $m = 1.5$ kg, the damping coefficient is $c = 0.8$ Ns/m and the damping and restoring forces are given in Tab.3 Eqn.(7) is solved using the Matlab built-in 4th order Runge-Kutta solver ODE45, for a several values of $\omega_e$. The principle of the Harmonic Balance (HB) can now be applied in order to compute the FRF. The HB is commonly applied to describe the structural dynamic behaviour of weak non-linearities and is based on the assumption that most of the response energy of a harmonically excited structure is concentrated in the frequency of excitation. In Fig. the FRF of the system with cubic stiffness is plotted when the amplitude of the excitation force is $F_0 = 0.2\text{N}$.

The results of the method described in Sec.2.1 are shown in Fig.3-5(b).

Firstly, the outcome of the nonlinear method applied to the linear system (using the same constants as the cubic hardening case with $k_{nl} = 0$) is plotted in Fig.3. As expected, the plots of the natural frequency, Eqn.(6a), and the loss factor, Eqn.(6b), as the amplitude of vibration changes, are constants.

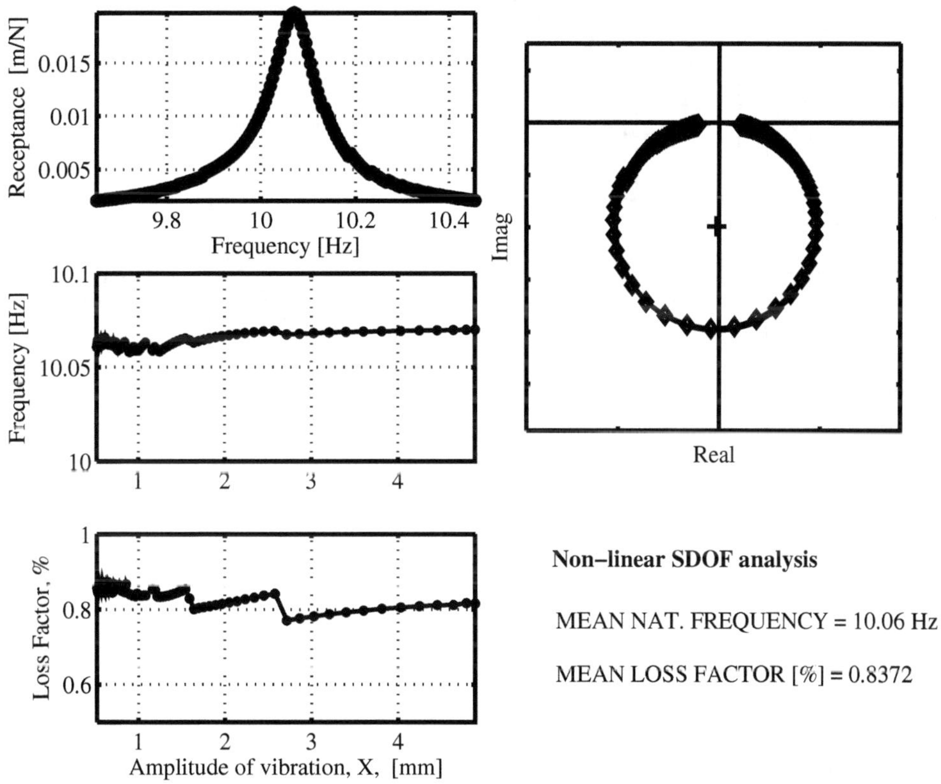


Fig. 3: Nonlinear analysis of the FRF of a linear system

On the contrary, Fig.4(a) and 4(b) depicts the nonlinear analysis of the FRF of the SDOF system with hardening cubic nonlinearity. Fig.4(a) shows that the natural frequency has a clear increasing trend with amplitude while the damping stays at a constant level. The characterisation analysis, Fig.4(b) provide instead the spatial parameters of the system: the modal mass (which in this case is the mass because the system has one degree of freedom) extracted is the same as the value used in the simulation. Also the damping coefficient agrees with the value used in the example. The top-right corner figure depicts the stiffness function: the circles are the data-points extracted and the solid line is the least-squares quadratic fit. In the example shown the quadratic function of the amplitude of vibration is

$$k(X) = k_1\,X + k_2\,X^2 + k_3\,X^3 \tag{8}$$

where $k_1 = 5946$, $k_2 = 6500$ N/m$^2$ and $k_3 = 2.7e^6$ N/m$^3$.

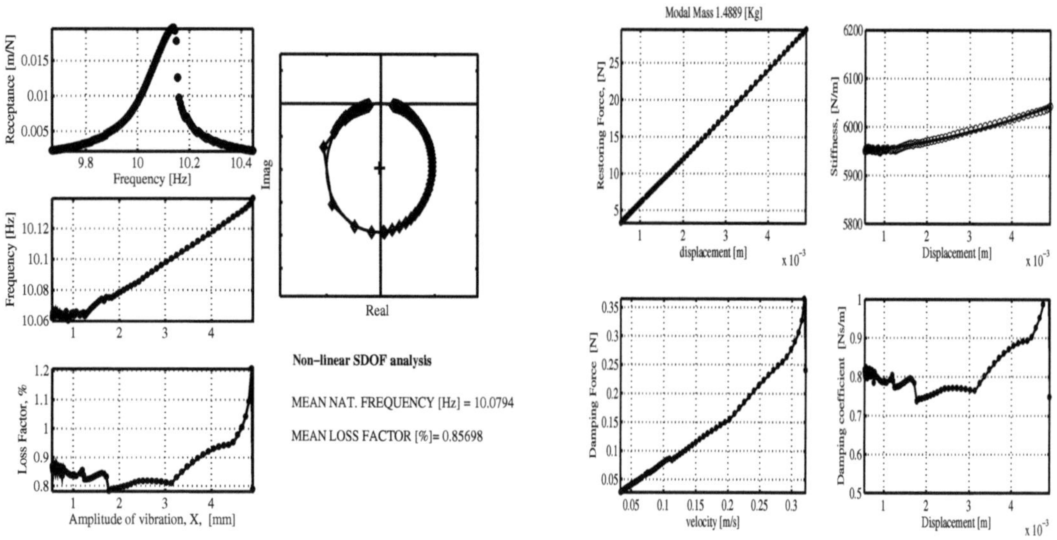


(a) (b)

Fig. 4: Nonlinear analysis of the FRF of a system with quadratic hardening stiffness: a) identification, b) characterisation

The results of nonlinear modal analysis on the nonlinear system with quadratic damping described above are shown in Fig.5(a) and 5(b). The identification procedure, Fig.5(a), reveals a constant natural frequency and a marked increase of loss factor with displacement which shows that the nonlinearities of this system lies in the damping rather than the stiffness. Looking at Fig.5(b) the plot of damping force versus velocity it can be seen that the extracted values (circles) can be interpolated, in the least-squares sense, with the quadratic function

$$c(\dot{X}) = c_1\,\dot{x} + c_2\,x^2 \tag{9}$$

where $c_1 = 1$ Ns/m and $c_2 = 4.76$ N(s/m)$^2$

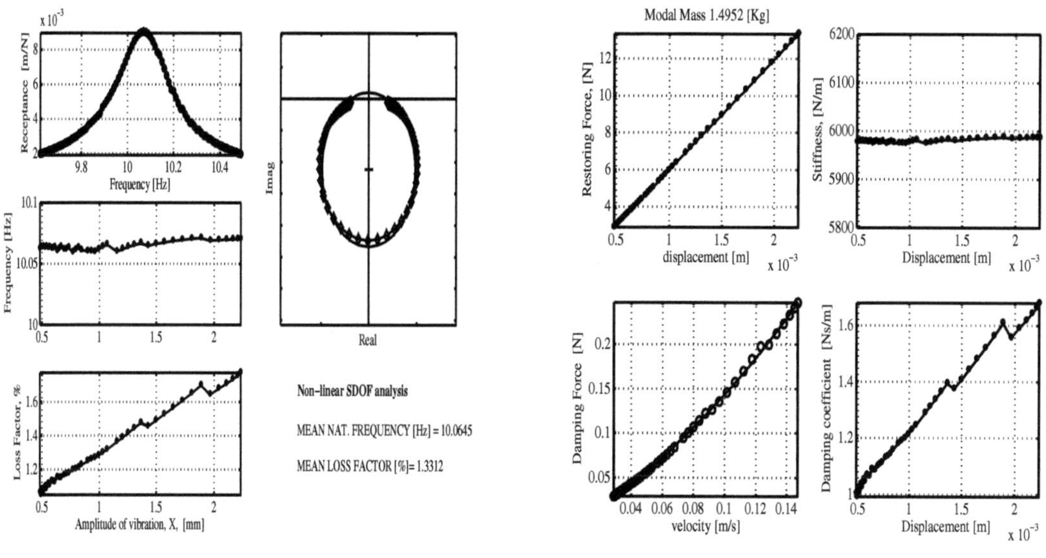


(a) (b)

Fig. 5: Nonlinear analysis of the FRF of a system with quadratic damping: a) identification, b) characterisation

# 4 Experimental data

The nonlinear modal analysis technique has been applied to the FRF measured on the test rig shown in Fig.6. The structure, also known as Nastran Tower, has been widely studied over the past 30 years at the Imperial College. It is known to exhibit a softening nonlinear stiffness. This is confirmed by the measured FRF shown in Fig.7.

The nonlinear modal analysis presented in Fig.8 is relative to the first mode as illustrative example. The top-right plot is obtained by overlaying the natural frequency vs amplitude of vibration for each force level: it can be seen that as the force is increased, so is displacement of the structure and the softening trend can be clearly observed. On the contrary, the bottom-right corner shows that the modal loss factor of the structure does not overlay when the force level is increased. Further investigation is necessary to identify what is the energy-dissipation mechanism and if/how this could be related to the amplitude of the excitation in addition to the amplitude of the response.

Fig. 6: Experimental setup. The rig represents a typical aerospace structure for the materials and method of assembly used (e.g. rivets)

## 5 CONCLUSIONS AND FUTURE WORKS

Nonlinearities in structural dynamics are a commonality more than a rarity. The major difficulty in obtaining reliable and precise numerical models is that these nonlinearities are not accounted for. In order to do so, there is the need of measuring, identifying, characterising and quantifying the nonlinear dynamic behaviour from experimental data. A review of the literature has revealed there the existing methods are not practical for implementation for large scale industrial applications. The method herein presented is based on the complex form of the frequency response function measured during standard vibration tests. Although it assumes that the response is at the same frequency as the excitation, i.e. the system is linearised, the analysis of numerical examples and measured data show that it is possible to identify and to characterise the nonlinearity to a first degree of approximation - which is already a significant

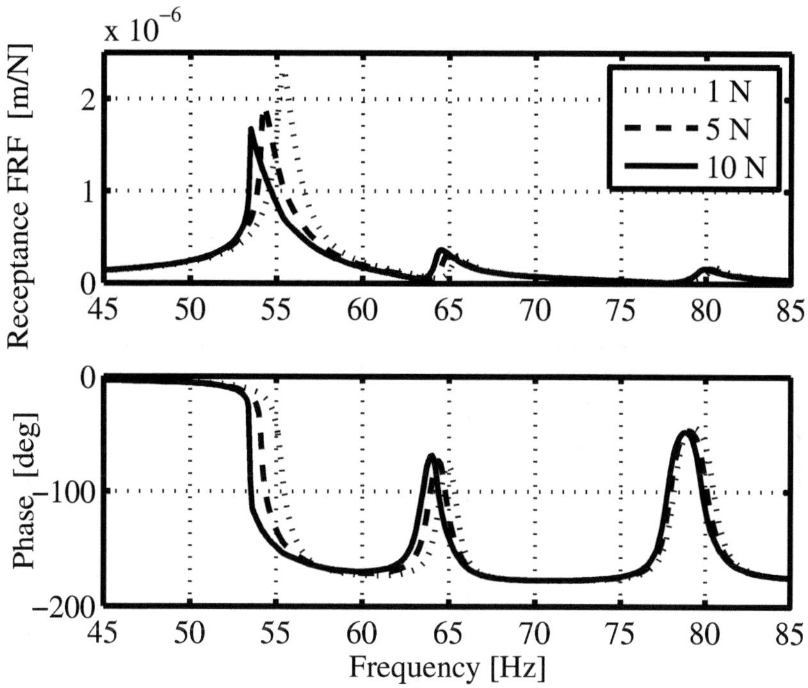


Fig. 7: Receptance FRF measured for different levels of excitation force

benefit to the modeller. Future work will focus on two different aspects: (i) evaluating to boundary of validity of the SDOF method, i.e. applying the method to MODF systems; (ii) using the extracted information for predicting the dynamic behaviour of a simple structure. First a numerical model will be built, e.g. FE model, then its response will be predicted using the nonlinear function computed with the nonlinear modal analysis and, finally, the results will be compared with experimental data.

## ACKNOWLEDGMENT

The first author expresses his gratitude to AgustaWestland for supporting this research.

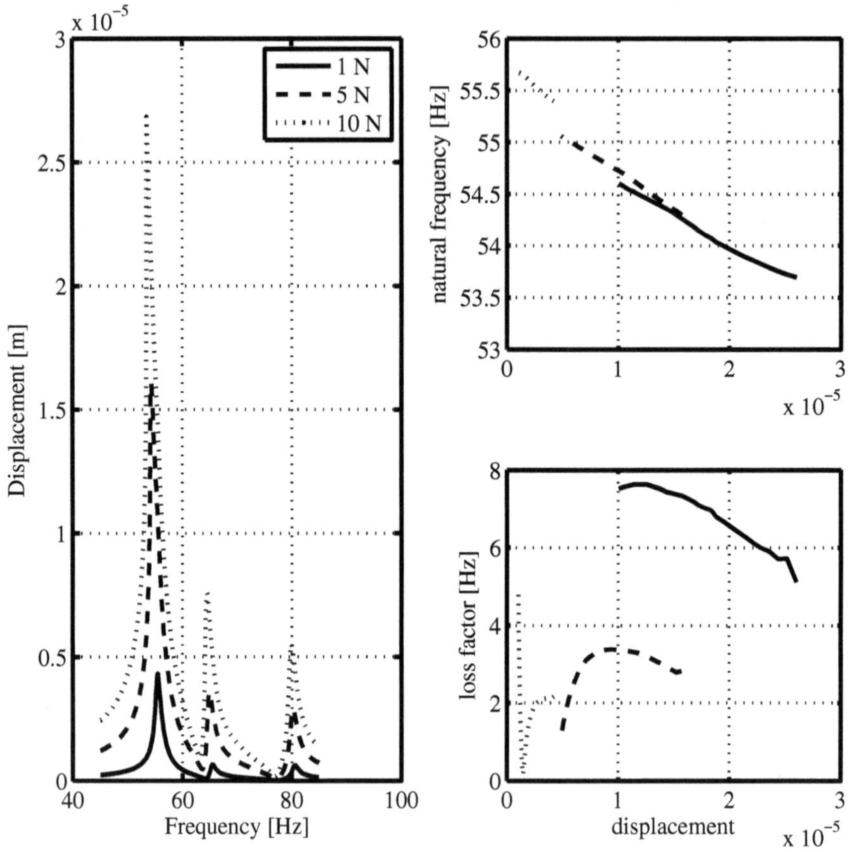


Fig. 8: Nonlinear modal analysis. The plot on the right of the figure show the change in natural frequency and loss factor with amplitude of vibration for the three different level of excitation

# References

[1] K. Worden and G.R. Tomlison. *Nonlinearity in Structural Dynamics.* Institute of Physics, 2001.

[2] G. Kerschen, K. Worden, A. F. Vakakis, and J. C. Golinval. Past, present and future of nonlinear system identification in structural dynamics. *Mechanical Systems And Signal Processing*, 20(3):505–592, April 2006.

[3] Dta handbook.

[4] S. F. Masri and T. K. Caughey. Nonparametric identification technique for non-linear dynamic problems. *Journal Of Applied Mechanics-Transactions Of The Asme*, 46(2):433–447, 1979.

[5] J. He and D.J. Ewins. A simple method of interpretation for the modal analysis of nonlinear structures. In SEM, editor, *IMAC V*, 1984.

[6] D.J. Ewins. *Modal Testing.* Research Studies Press, UK, 2000.

[7] D. Göge, M. Sinapius, U. Fullekrug, and M. Link. Detection and description of non-linear phenomena in experimental modal analysis via linearity plots. *International Journal Of Non-Linear Mechanics*, 40(1):27–48, January 2005.

[8] D. Göge, U. Fullekrug, M. Sinapius, M. Link, and L. Gaul. Advanced test strategy for identification and characterization of nonlinearities of aerospace structures. *Aiaa Journal*, 43(5):974–986, May 2005.

[9] D. Göge and J. M. Sinapius. Experiences with dynamic load simulation by means of modal forces in the presence of structural non-linearities. *Aerospace Science And Technology*, 10(5):411–419, July 2006.

[10] D. Göge. Fast identification and characterization of nonlinearities in experimental modal analysis of large aircraft. *Journal Of Aircraft*, 44(2):399–409, March 2007.

[11] Sinapius M. Gloth G. Influence and characterisation of weak non-linearities in swept-sine modal testing. *Aerospace Science and Technology*, 8:111–120, 2004.

[12] Sinapius M. Gloth G. Swept-sine excitation during modal identification of large aerospace structures. Technical Report Forschungsbericht 2002-18, Gernan Aerospace Center, DLR, Institute of Aeroelasticity, 2002.

[13] R. Lin. *Identification of the dynamic characteristic of nonlinear structures.* PhD thesis, Dept Mechanical Eng. - Imperial College - London, 1990.

Proceedings of the IMAC-XXVIII
February 1–4, 2010, Jacksonville, Florida USA


# Detection and Quantification of Nonlinear Dynamic Behaviors in Space Structures

**Aurélien HOT[(1)], Gaetan KERSCHEN[(2)], Emmanuel FOLTÊTE[(1)], Scott COGAN[(1)], Fabrice BUFFE[(3)], Jerôme BUFFE[(4)], Stéphanie BEHAR[(4)]**

[(1)] FEMTO-ST Institute - Applied Mechanics Department, 24 rue de l'Epitaphe 25000 Besançon, France
aurelien.hot@univ-fcomte.fr, scott.cogan@univ-fcomte.fr, emmanuel.foltete@univ-fcomte.fr

[(2)] Department of Aerospace and Mechanical Engineering, University of Liege, Chemin des Chevreuils 1 (B52), 4000 Liège, Belgium
g.kerschen@ulg.ac.be

[(3)] Centre National d'Etudes Spatiales, 18 Avenue Edouard Belin 31401 Toulouse Cedex 9
fabrice.buffe@cnes.fr

[(4)] Thales Alenia Space, 100bd du Midi - BP99 - 06156 Cannes la Bocca Cedex
jerome.buffe@thalesaleniaspace.com, stephanie.behar@thalesaleniaspace.com

## ABSTRACT

Following a campaign of structural dynamic measurements on an industrial structure, the question often arises: "Is my structure non-linear?". A response to this question is important to the extent that the presence of a non-linearity, even local, can significantly affect the global dynamic behavior of a structure. Several techniques that enable engineers to detect a non-linear behavior can be found in the literature. These methods are applied mostly in the frequency domain and give the best results with a stepped sine excitation.
The goal of this paper is to propose an alternative methodology. It is based on the principal component analysis and uses time responses obtained with a random excitation. This will be first applied to an academic simulated system, and then tests are carried out on a simplified solar array system.

## 1. INTRODUCTION

Structural dynamic behavior must generally be taken into account in the design of mechanical systems in order to insure their performance and reliability. Confrontation between numerical simulations and experimental observations on a prototype often indicates that the model is unsatisfactory and model updating methodologies have been developed over the years to support the engineer in improving model quality. However, most of these strategies are restricted to linear dynamic models. Indeed, the lack of knowledge and tools in the field of non-linear structural dynamics, the complexity of non-linear models and the long duration of calculations have for a long time put off engineers, even if most industrial structures show some degree of non-linearity. Hence, new approaches for validating non-linear systems in structural dynamics must be developed.
Non-linear system identification is an integral part of the validation process and it can be viewed as a succession of three steps: detection, characterization and parameter estimation. The first step enables to detect a non-linear behavior on the tested structure while the characterization step localizes and determines the type and the functional form of the non-linearity. Finally, the parameter estimation step evaluates the coefficients of the non-linear model. This paper will focus on the detection step.
Several methods that have been proved useful for detecting non-linearity in structural responses can be found in the literature. The most well known techniques use data from the frequency domain, for example: the

homogeneity test that examines distortions in the Frequency Response Functions (FRF); the detection of jump phenomenon in the FRF due the non uniqueness of the solutions for a non-linear system; the Hilbert transform that differs from the original FRF. Readers can refer to [1] and [2] for more details on these detection methods. Moreover, the results of these methodologies depend on the excitation type. The stepped-sine excitation is the best one since it gives a well defined FRF and the distortion appears clearly. But, due to its high cost, the most commonly used excitation type is random or sweep-sine.

Fewer techniques exist using only time data. For example, in [3] the Hilbert transform is applied to signals in the time domain in order to extract the instantaneous dynamic characteristics of the structure. Then, using the time varying envelope as well as the instantaneous phase and frequency, the authors are able to detect a non-linear behavior. The continuous wavelet transform used in [4] is a method that uses the free response of a non-linear system. The authors detect non-linearities by looking at distortions in the amplitude and the phase of the wavelet.

The aim of this study is to apply a methodology based on the principal component analysis in order to detect a non-linear behavior. This approach has two main advantages. First, it uses only time domain data so that no signal processing transformation of the measurements is needed and all the information is conserved. Second, the type of excitation used here is random, which is generally not chosen because of its poor results in the frequency domain.

## 2. PRINCIPAL COMPONENT ANALYSIS

### 2.1 Theory

Principal Component Analysis (PCA) is a statistical multivariate analysis technique whose goal is to reduce the dimension of a response matrix $X$ and to retain the dominant information. PCA is closely related to the Singular Value Decomposition (SVD) and the Proper Orthogonal Decomposition, also known as the Karhunen-Loève decomposition.

One specific application of the PCA in the field of structural dynamics is to find the subspaces spanned by the principal directions that contain most of the system's energy, without calculating the modes shapes.

Given a response matrix $X$ containing the displacements $x_i(t_j)$ obtained over n sensors on the studied structure during m time samples:

$$X = \begin{pmatrix} x_1(t_1) & \cdots & x_1(t_m) \\ \vdots & \ddots & \vdots \\ x_n(t_1) & \cdots & x_n(t_m) \end{pmatrix} \tag{1}$$

The SVD is then calculated in order to obtain the principal directions:

$$X = U\Sigma V^T \tag{2}$$

where: - $U$ is an (n x n) orthonormal matrix containing the left singular vectors of $X$ which are the principal directions, also known as the Proper Orthogonal Modes (POM);

- $\Sigma$ is an (n x m) pseudo-diagonal matrix whose main diagonal contains the singular values of $X$, also known as the Proper Orthogonal Values (POV) and which correspond to the energy contained in the POM;

- $V$ is an (m x m) orthonormal matrix containing the right singular vectors of $X$ (the time history of the response).

In structural analysis, the idea of using the subspaces spanned by a subset of vectors contained in $U$ has already been employed in the past to detect and localize structural damage [5] or to perform detection, isolation and reconstruction of faulty sensors [6].

In this paper, the subspaces will be compared for different levels of excitation. Because of the homogeneity principle, if the system is linear, the subspaces should be identical. However, for a non-linear system, as the level of excitation increases, the subspace spanned by the retained vectors in $U$ becomes more and more distorted. These distortions are quantified in the following sections.

## 2.2 Application to non-linear detection

### 2.2.1 Angle between subspaces

One means to compare two different subspaces is to measure the principal angle $\theta$ between them. Let $X_L$ be the linear reference data matrix, defined by equation (1) and obtained for a low level of excitation. Let $X_{NL}$ be data matrix obtained for a high level of excitation with potential non-linear distortion effects. Their singular value decomposition gives:

$$X_L = U_L \Sigma_L V_L{}^T \tag{3}$$

$$X_{NL} = U_{NL} \Sigma_{NL} V_{NL}{}^T \tag{4}$$

In order to retain only the $q_L$ and $q_{NL}$ principal components of respectively $U_L$ and $U_{NL}$, these matrices are partitioned into two parts:

$$U_L = (U_L^1 \quad U_L^2) \tag{5}$$

$$U_{NL} = (U_{NL}^1 \quad U_{NL}^2) \tag{6}$$

where $U_L^1 \in \Re^{n,q_L}$ and $U_{NL}^1 \in \Re^{n,q_{NL}}$ containing only the vectors related to the main singular values, for example containing more than 98% of the system total energy, and $U_L^2 \in \Re^{n,n-q_L}$ and $U_{NL}^2 \in \Re^{n,n-q_{NL}}$ the matrix regrouping the remaining vectors.
Then, the different angles $\theta_i$ are calculated between the subspaces $U_L^1$ and $U_{NL}^1$ [7]:

$$\theta_i = \left\langle U_L^1, U_{NL}^1 \right\rangle \quad \forall i \in [1, q] \tag{7}$$

The angle $\theta$ retained to quantify the subspace distortion is the largest value of $\theta_i$.

In theory, if the subspaces are identical then $\theta$ is equal to 0. The more the subspaces differ due to increasing non-linearity, the more $\theta$ tends towards 90°. However, in practice, even if th e two subspaces are identical $\theta$ will not be equal to zero. This is due to the random excitation and to the noise inherent in experimental measurements.
In order to avoid these random phenomena and to be sure that the calculated $\theta$ is due to non-linear distortion, a limit of linearity ($LL$) is calculated. First, the time response of the reference configuration is partitioned into $p$ different parts, indicated by the red dashed windows (see Fig.1). Then $p$ values of $\theta$ are calculated between the subspace spanned by the whole signal and each of these parts. Finally, the limit of linearity is defined as the mean value plus three times the standard deviation of the $p$ values of $\theta$:

$$LL = \bar{\theta} + 3.\sigma_\theta \tag{8}$$

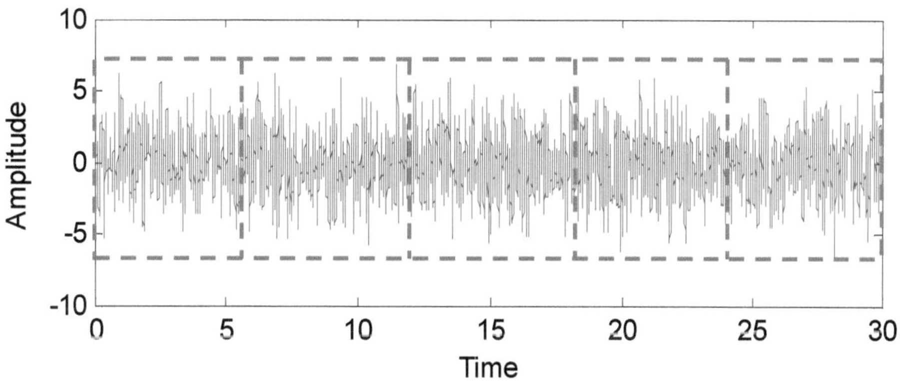


Figure 1: Windowing used to calculate the limit of linearity

Now comparing two subspaces from a same system but at two different levels of excitation, this definition will guarantee, within a 99.7% confidence interval, that if $\theta$ is above $LL$ then the system is non-linear.

### 2.2.2 Subspace residual projection

The second criterion developed in this paper is based on the calculation of the residual error $r$ obtained after projection of the first subspace onto the second.
Let $U_L$ be the reference subspace spanned by data at low level of excitation and let $x_{NL}(t)$ be the time vector of the non-linear response at $t$ time. The projection vector $x^*_{NL}(t)$ of $x_{NL}(t)$ on $U_L$ is defined as (9).

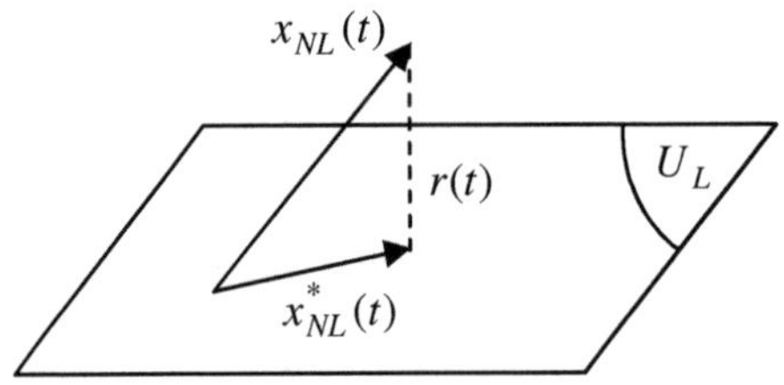


Figure 2: Geometrical view of the residual error criterion

$$x^*_{NL}(t) = U_L U_L^T x_{NL}(t) \tag{9}$$

Thus, for each time step, the residual error $r(t)$ is given by (10). The value $r$ retained as a criterion in this study is defined as the maximum value of $r(t)$.

$$r(t) = \left\| x^*_{NL}(t) - x_{NL}(t) \right\| \tag{10}$$

$$r = \max_t r(t) \quad \forall t \in [1;m] \tag{11}$$

The same kind of limit of linearity defined in previous section is calculated. In this case, a residual error value is calculated for each part in order to find the value of $LL$. Thus, the more $r$ is large and above $LL$, the more the dynamic behavior is non-linear.

## 3. NUMERICAL APPLICATION

### 3.1 Study case

The tested structure is a beam clamped at one end. Between the free end and the ground, a non-linear cubic spring Knl is first added. In a second example this spring is replaced by a quadratic damper Cnl. All of the characteristics are provided in Table 1 and the first eigenfrequencies of the linear beam (without the non-linear elements) are given in Table 2.

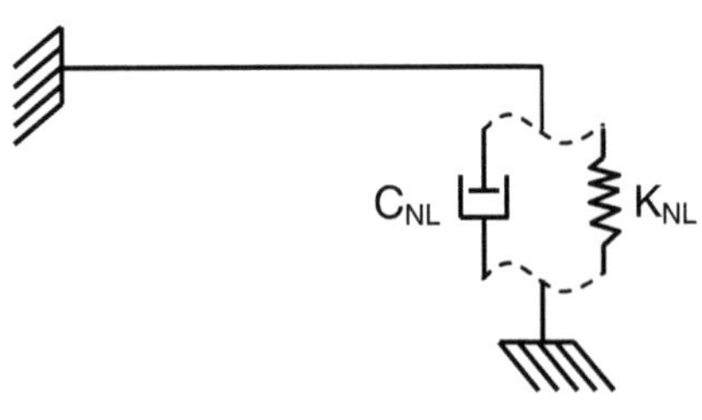


Figure 3: Two configurations of the studied beam

Table 1: Characteristics of the studied system

| Length (m) | Width (m) | Thickness (m) | Number of elements | E (Pa) | ρ (kg.m$^{-3}$) | Knl (N.m$^{-3}$) | Cnl (N.s$^2$.m$^{-2}$) |
|---|---|---|---|---|---|---|---|
| 1 | 0.014 | 0.014 | 10 | 2.1$^e$11 | 7800 | 1e7 | 100 |

Table 2: Eigenfrequencies of the linear beam

| Mode | 1 | 2 | 3 | 4 | 5 |
|---|---|---|---|---|---|
| Frequency (Hz) | 11.73 | 73.54 | 205.96 | 403.87 | 668.67 |

The beam is excited with a constant amplitude random noise in the frequency band [1 Hz – 250 Hz]. In order to compare several levels of non-linearity, six different input amplitudes are tested? The first one is at a very low level of 0.001 NRMS in order to guarantee a linear behavior, further referred to as the reference data. The others correspond to amplitudes of 0.5, 1, 1.5, 2 and 2.5 NRMS where non-linear behaviors are expected.

For the simulations, the number of vectors retained in the matrix $U$ is equal to 3. Indeed, these three first POM concentrate more than 98% of the total energy of the system.

### 3.2 Results

In order to calculate the principal angle between subspaces and the residual projection error, the time responses are first divided into 10 parts. Then $\theta$ and r are calculated using one part at a time and the whole reference data. Thus 10 values of $\theta$ and r, corresponding to each part of the signal, are calculated for all excitation levels. This process allows isolating, in the time domain, a potential non-linear behavior.

The results are reported Figure 4 and 5. The legend defined in Table 3 should be referred to for all graphics in this section. Looking at the shape of $\theta$, the first level for which it exceeds the limit of linearity is 1NRMS and it does so for the part number 6. For higher levels, $\theta$ exceeds $LL$ more clearly and for a more important number of parts. Now regarding the residual projection error, all the values are far above the limit of linearity.

So for both cases, the values of the criteria increase with the input level and thus with the level of non-linearity. The residual projection error seems however much more sensitive to a change in the system behavior.

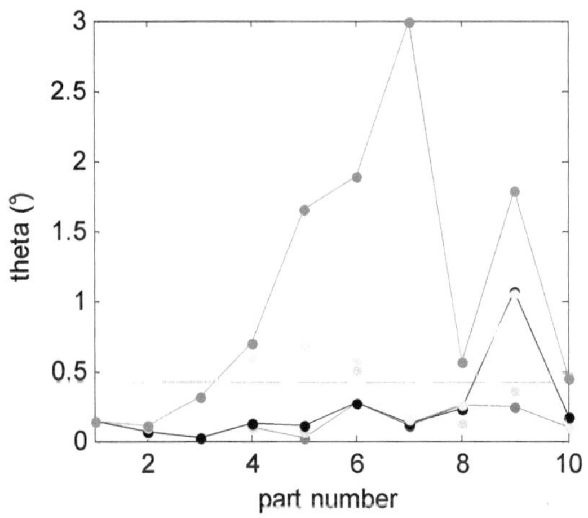


Figure 4: Evolution of $\theta$ for a non-linear cubic stiffness system

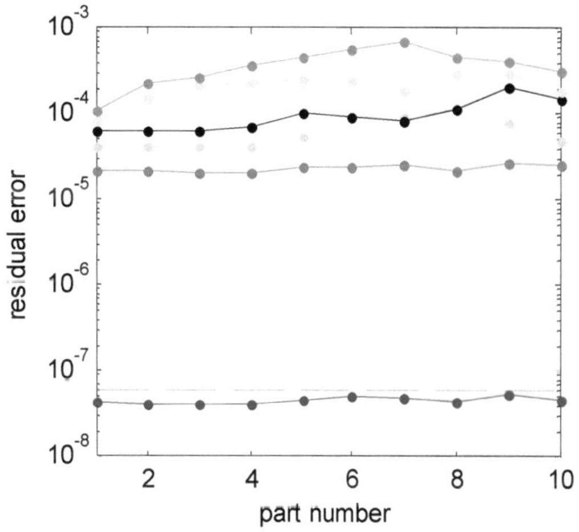


Figure 5: Evolution of r for a non-linear cubic stiffness system

Table 3: Graphics legend

| | | | |
|---|---|---|---|
| —— | 0.001 NRMS | —— | 2 NRMS |
| —— | 0.5 NRMS | —— | 2.5 NRMS |
| —— | 1 NRMS | | |
| —— | 1.5 NRMS | ---- | $LL$ |

The frequency responses are then computed (Fig.6) in order to compare the results with the proposed criteria. These FRF become clearly distorted after an excitation level of 1 NRMS: the first mode tends to shift to higher frequencies and the distortions observed around 45 Hz and 105 Hz correspond to odd harmonics of this first mode. These observations are in agreement with the results of $\theta$ and r given previously.

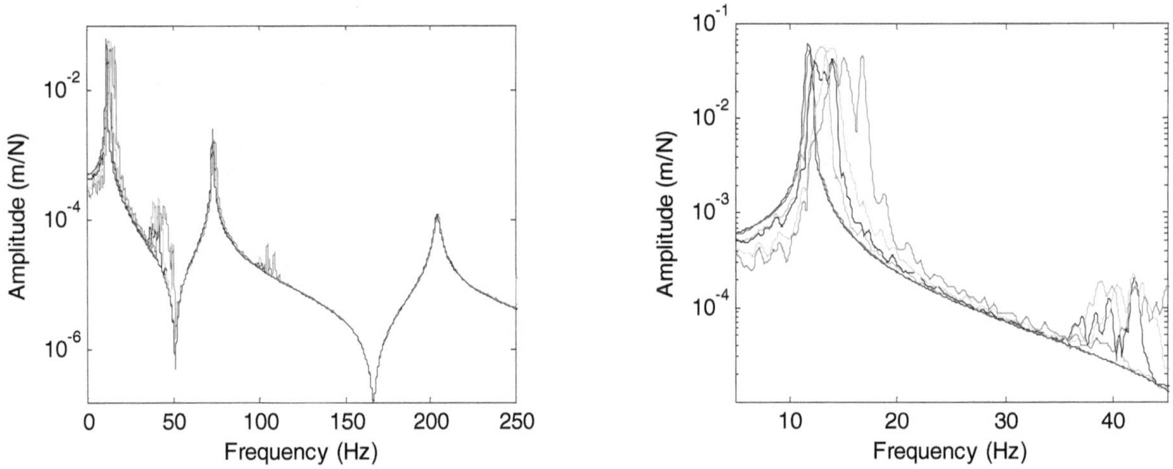


Figure 6: FRF of the non-linear cubic stiffness system

In this second example, the same study is carried out on the same beam but with a quadric damper. On the FRF (Fig.7), the influence of an increasing value of the force is characterized by a reduction in the magnitude of the resonance peak. Regarding the evolution of $\theta$ and r (Fig.8 and Fig.9), the remarks are analogous to those in the non-linear stiffness example: their values increase with the input excitation and the residual projection error proves to be far more sensitive.

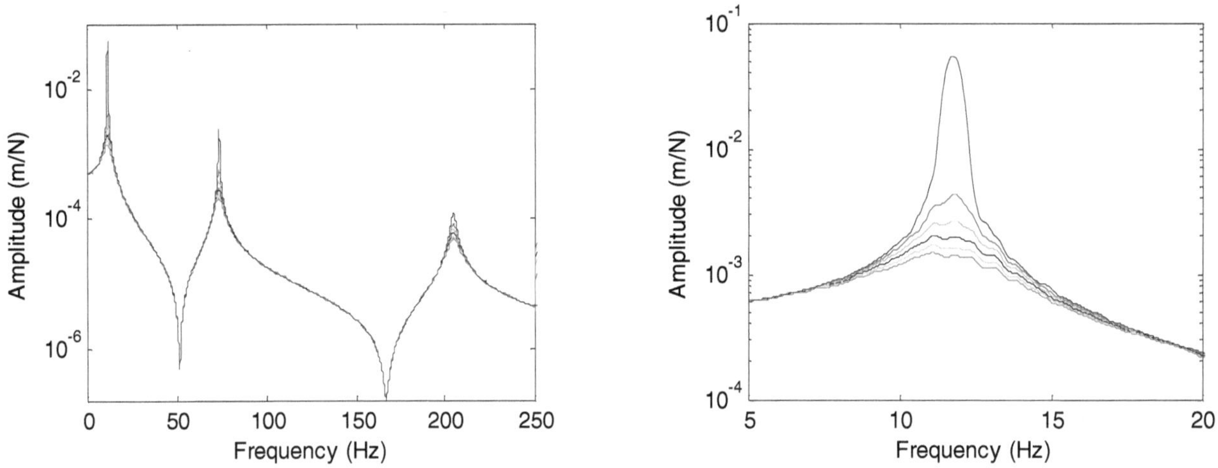


Figure 7: FRF of the non-linear quadratic damper system

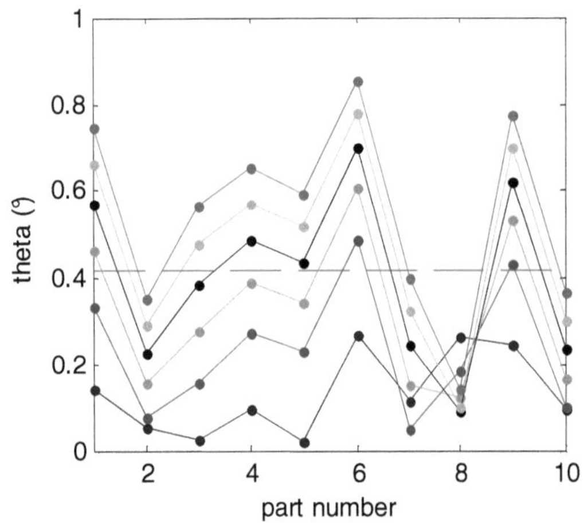


Figure 8: Evolution of $\theta$ for a non-linear quadratic damper system

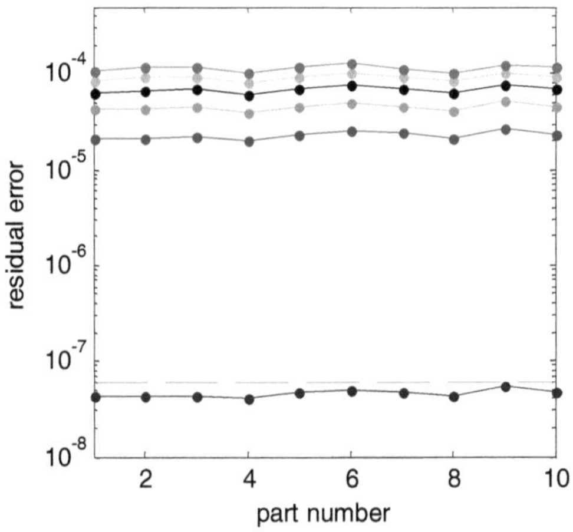


Figure 9: Evolution of r for a non-linear quadratic damper system

In conclusion, the two proposed criteria enable to detect a non-linear behavior on a simple structure using only the time responses. Looking only at the FRF distortions could be enough for detecting non-linearities but the concept of limit of linearity proposed here limits subjective decisions. The criteria also allow the level of non-linearity to be quantified by comparing their values at low and high levels of excitation.

### 3.3 Influence of noise

In section 2.2.1 the question of the noise has arisen. Indeed, during measurement campaigns, multiple sources of noise can occur such as a badly fixed sensor or a defective cable. It is thus important to insure that the two criteria are only sensitive to physical non-linear behaviors and that they do not exceed the limit of linearity due to the presence of random noise.
To investigate this question, a random noise is added to the measurement results $X$ obtained for the low level of excitation (where the system is supposed to be linear). A 1, 2, 5, and 10% noise level $nl$ is successively added to $X$ and gives the new data $X_{noise}$ defined by (12), $rand$ is a random value in the interval [-1;1].

$$X_{noise} = X(1 + nl * rand) \tag{12}$$

The FRF obtained for the different noise levels (Fig.10) show the clear influence of the noise on the measurements with respect to the exact FRF. However, looking at Figure 11 the criterion based on the angle between subspaces is not affected by the noise. Indeed, since only the principal components of the signal are kept after the SVD, all the noise is not taken into account during the calculation of $\theta$. Regarding the second criterion (Fig.12), the value of r varies slightly, but even for large distortions of the FRF, r does not exceed the limit of linearity.
These results confirm that the variations of $\theta$ and r, observed in the previous section are indeed due to physical non-linear effects are not to the kind of noise modeled here.

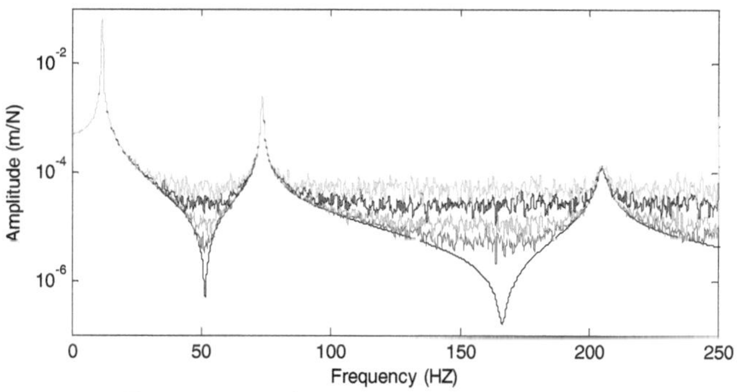


Figure 10: Influence of noise on the FRF
— without noise; — 1%; — 2%; — 5%; — 10%

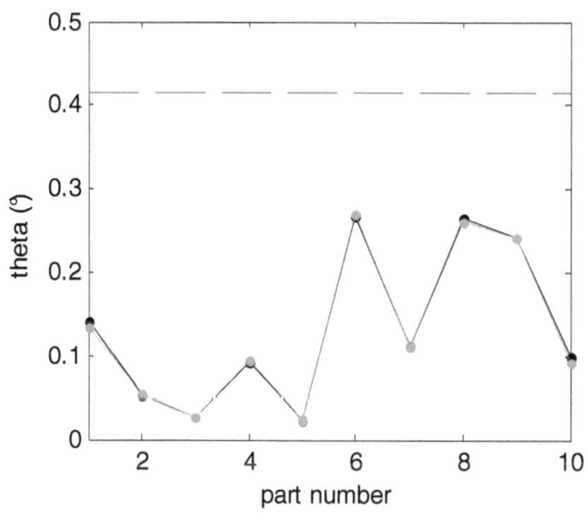


Figure 11: Evolution of $\theta$ for data with noise
— without noise; —1%; —2%; — 5%; — 10%; - - *LL*

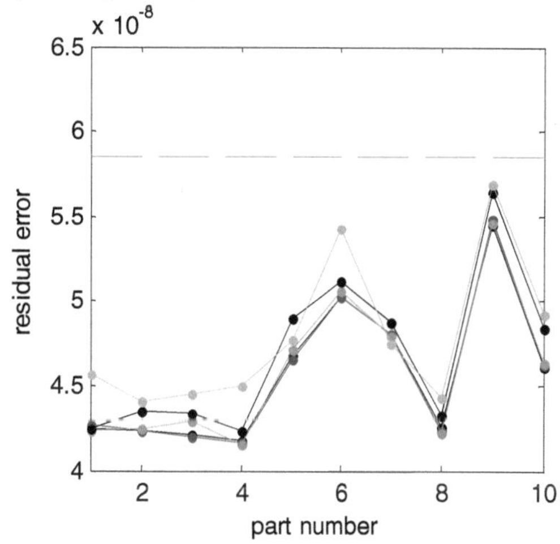


Figure 12: Evolution of r for data with noise
— without noise; — 1%; —2%; — 5%; — 10%; - - *LL*

## 4. EXPERIMENTAL APPLICATION

### 4.1 Dual plate system

A specific difficulty encountered on space structures is the behavior of solar generators in their stored position. These solar panels, during the spacecraft launch, are folded and impact one another through non-linear material blocks. Boundary conditions are also not well controlled and gaps may appear in clamping as the excitation level of the structure increases. These are just some of the many conditions which lead to non-linear behaviors.
In order to study this specific case, a simplified experimental test rig has been designed (Fig.13).

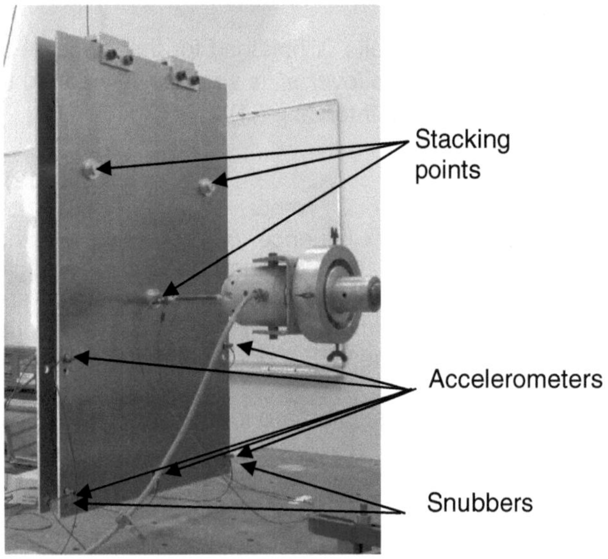


Figure 13: Experimental set-up

It is composed of two aluminum plates clamped together on one edge (at the top on the photo) and through three stacking points. The system is mounted in a free-free configuration. It is instrumented with five accelerometers on the free edges of each plate and with an impedance head (accelerometer + force sensor) at the excitation point.
The shaker generates a random signal with constant amplitude from 5 Hz to 312 Hz at several levels of amplitude in order to activate more or less the non-linearities.
Two types of non-linearities are present in the structure. First, the imperfect tightening of bolts can lead to clearance in the joints. Moreover, the small contact surface between the stacking points and the plates can introduce large bending phenomena concentrated in these areas. Second, impacts are introduced by adding two steel snubbers on the two corners of the free edge. A 1 mm clearance is maintained between the snubber and the plate, such that there is no contact at rest.

The goal here is to confirm that the two criteria defined previously can be an alternative to usual techniques in the field of detection of non-linear behavior.

### 4.2 Results

The frequency responses of this structure are given Figure 14. The FRF are slightly affected at 0.37 NRMS and clear distortions appear after a level of 0.74 NRMS. For the modes at 75, 177 and 252 Hz, the peak amplitudes are strongly decreased because of the influence of impacts.

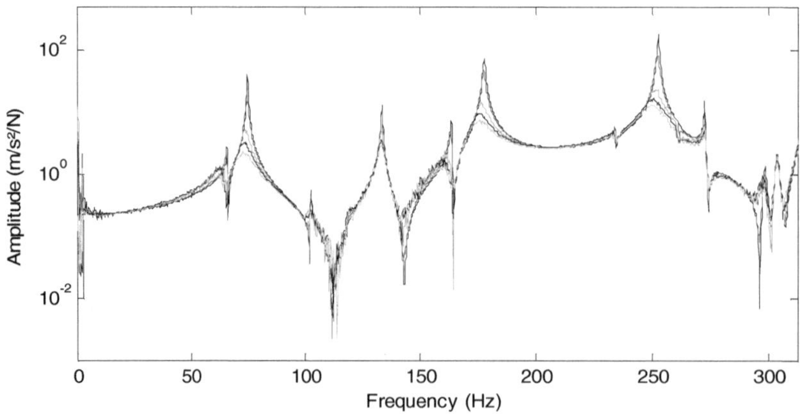


Figure 14: FRF of the experimental set-up for several excitation levels
—0.15 NRMS; — 0.37 NRMS; — 0.74 NRMS; — 1.14 NRMS; — 1.60 NRMS

The distortions observed on the FRF are confirmed by the two proposed criteria. The one based on the angle between subspaces (Fig.15) detect a non-linear behavior after 0.74 NRMS. The other one calculating the residual projection error (Fig.16) is, as for the simulated beam, more sensitive and exceeds the limit of linearity for a lower level.

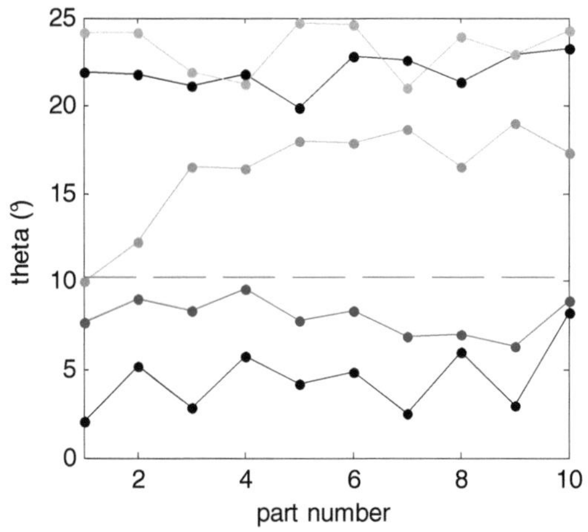


Figure 15: Evolution of $\theta$ for the experimental set-up
—0.15 NRMS; — 0.37 NRMS; — 0.74 NRMS;
— 1.14 NRMS; — 1.60 NRMS; - - *LL*

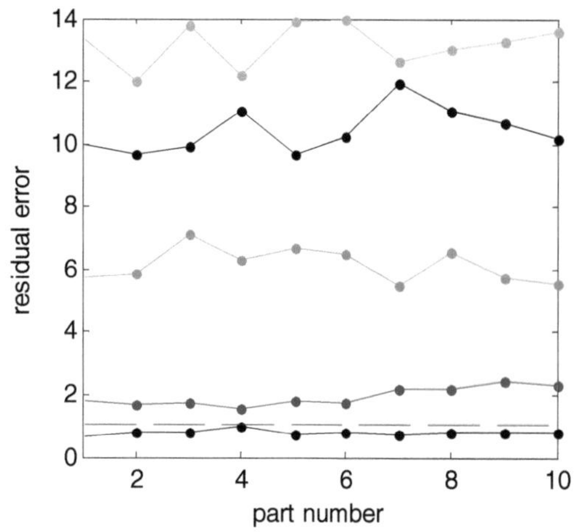


Figure 16: Evolution of r for the experimental set-up
— 0.15 NRMS; — 0.37 NRMS; — 0.74 NRMS;
— 1.14 NRMS; — 1.60 NRMS; - - *LL*

## 5. CONCLUSION

This paper provides two alternative methodologies in the field of non-linear detection. Using only time responses and after a singular value decomposition, the principal angle and a residual projection error between sub-spaces can be calculated. The results obtained in this study show that these two criteria appear to be useful tools for detecting and quantifying non-linear behavior on a structure. The principal angle criterion seems more suitable for high levels of non-linearities but is insensitive to noise on measurements. The residual projection error detects lower levels of non-linear behavior but is more sensitive to noise. However, the notion of a limit of linearity helps to differentiate the sources of the observed perturbation.
These two nonlinearity indicators have been applied here following a campaign of experimental tests, but they could also be applied during the test. This provides a real-time indicator that can support the engineers in improving the knowledge of their structure. Finally, future work will incorporate these criteria into a more global decision-making indicator allowing the impact of the non-linear effects to be weighed in light of the design decisions to be made.

## 6. REFERENCES


[1] G. Kerschen, K. Worden, A.F. Wakakis, J.C. Golinval, Past, present and future of non-linear system identification in structural dynamics, Mechanical Systems and Signal Processing 20 505-592 (2006).
[2] K. Worden, G.R. Tomlinson, Nonlinearity in structural dynamics, Detection, Identification and Modelling, IoP (2001).
[3] M. Feldman, Considering high harmonics for identification of non-linear systems by Hilbert transform, Mechanical Systems and Signal Processing 21 943-958 (2007).
[4] M.N. Ta, J. Lardiès, Identification of weak nonlinearities on damping and stiffness by the continuous wavelet transform, Journal of sound and vibration 293 16-37 (2006).
[5] P. De Boe, J.C. Golinval, Principal component analysis of a piezosensor array for damage localization, Structural health monitoring 2 137-144 (2003).
[6] G. Kerschen, P. De Boe, J.C. Golinval, K. Worden, Sensor validation using principal component analysis, Smart Materials and Structures 14 36-42 (2005).
[7] A. Bjorck, G.H. Golub, Numerical methods for computing angles between linear subspaces, Mathematics of computation 27 579-94 (1973).

Proceedings of the IMAC-XXVIII
February 1–4, 2010, Jacksonville, Florida USA


# An Approach to Non-linear Experimental Modal Analysis

Michael Link[1)], Marc Boeswald[2)], Sebastien Laborde[3)], Matthias Weiland[1)], Adriano Calvi[4)]

[1)] University of Kassel, Germany, [2)] DLR Institute of Aeroelasticity, Goettingen, Germany
[3)] EADS Astrium, Toulouse, France, [4)] ESA, Noordwijk, The Netherlands

## ABSTRACT

It was the aim of the DYNAMITED (DYNamics AssessMent and Improvment of TEst Data) research project conducted by a European industrial and university team associated around EADS ASTRIUM and funded by the European Space Agency (ESA) to assess and improve dynamic test procedures. In the field of post test activities, the detection, identification, quantification and prediction of the non-linear behavior of space structures is of prime importance. Present evaluation methods for spacecraft vibration tests are based on the assumption of linear structural behaviour. The resonance shifts and FRF peak variations observed in the case of non-linear structural behavior are generally not reflected in practice by non-linear evaluation procedures. In order to avoid overloading of the structure during the qualification test on a shaking table the dynamic response is generally controlled at specified levels and locations by input notching. This approach generates an effectively quasi linear structural behavior at the different input levels which enables the utilization of classical linear modal extraction tools to be applied separately at each level. However, the measured dynamic responses (transmissibilities) reveal peak shifts and amplitude changes depending on the input level of the base excitation.
In the paper we present an approach using three different input levels where the response levels are controlled to be constant within a narrow band frequency range around the dominant resonances. We describe the premises of a technique with the aim of predicting the responses to ***other than the measured*** input levels. This is achieved by applying interpolation and extrapolation techniques to the modal data extracted from the transmissibilities measured at three different input levels. Results are presented from an application of the technique to the vibration test data measured during a typical STM satellite structure (Structural Test Model) test campaign.

## 1. INTRODUCTION

In space industry present experimental vibration test evaluation methods are generally based on the assumption of linear structural behaviour. Test procedures to check the linearity assumption are described in the European Space Agency (ESA) standard on modal survey assessment (ECSS-E-30-11A) and are applied for base driven tests on the shaking table as well as in modal survey testing. The resonance shifts and FRF peak variations observed in the case of non-linear structural behaviour are generally not reflected in practice by non-linear analytical modelling. Instead equivalent linear modelling adapted to a specified load level is used.
This approach can be tolerated since experience shows that in many practical cases the non-linear structural behaviour is not too strong so that the experimental modal data still lie within the natural scatter generated by other sources of test data variability like fabrication tolerances, multiple assembly or test reproducibility.

When evaluating test data obtained from spacecraft (S/C) qualification testing on a shaking table it is an important goal to characterise the non-linear behaviour of the structure with respect to its magnitude (weak or strong) to enable a decision if equivalent linear modelling is applicable or not. Since existing codes for experimental modal analysis (EMA) rely on the linearity assumption, the scatter of the modal identification results when applying such codes to non-linear vibration test data will be particularly high. It was therefore one of the goals of a research project realized by a European industrial and university team associated around EADS ASTRIUM and funded by the European Space Agency (ESA) to identify non-linear modal parameters in addition to the traditional modal parameters related to the underlying linear system.

## 2. DETECTION OF NON-LINEAR RESPONSE PROPERTIES

A large number of methods are described in the literature directed towards detection and quantification of structural non-linearities (e.g. refs.[1] – [6]). One of the best ways of dealing with non-linearities in practical structures is to control the vibration levels during the measurements. For example, Figure 1(a) shows the frequency response functions (FRF) of an analytical single-DoF system with a cubic stiffness non-linearity calculated under sinusoidal excitation force with different constant amplitudes. It can be seen that the distortion of the FRFs is increasing with increasing force level. As a result, problems can be expected when applying classical (linear) experimental modal analysis to such deformed FRFs. The modal parameters extracted this way will suffer from inaccuracy, especially for the damping and the modal mass due to the distortion of the response curve. However, the analysis of FRF distortions can be used for the characterization of the non-linearity, e.g. by comparison with analytical non-linear FRFs of different types of non-linearities, e.g. ref. [1]. Keeping the excitation force level constant would mean to increase the level of the drive signal of the electro-dynamic shaker when approaching resonance. This involves an unavoidable risk of damaging the structure under test, especially in cases of lightly damped structures and powerful shakers which normally is not accepted in real industrial test procedures.

Figure 1(b) shows the calculated FRFs of the same non-linear system now keeping the response amplitude constant. The legend in Figure 1(b) shows the levels of the constant displacement responses of the different simulations. In contrast to Figure 1(a) it can be seen that the excitation force was allowed to change so that the response was maintained at a constant level. By keeping the response constant, the non-linearity was also kept constant and thus each FRF in Figure 1(b) reflects a linear characteristics (not distorted) with slightly different underlying linear stiffness (in contrast to Figure 1(a) where all FRFs exhibit non-linear distortions).This approach generates an effectively quasi linear structural behaviour around the main resonances at the different input levels which enables the utilization of classical linear modal extraction tools to be applied separately at each response level. However, the measured dynamic responses (transmissibilities) reveal peak shifts and amplitude changes depending on the input level of the base excitation which can be utilized to describe the non- linear behaviour.

Another advantage of the constant response level testing is that the structure under test can be prevented from being damaged. This requires that the response levels to be investigated have to be chosen carefully. One of the drawbacks of this method is that it is only applicable in a narrow frequency band around the resonance of an isolated non-linear mode.

In industrial qualification procedures for spacecraft testing it is common practice to apply input notching in order to avoid overloading of the structure during the qualification test on a shaking table which requires the control of the dynamic response at specified levels and locations. In this case the table acceleration levels are reduced in the range of the main resonances such that the structure's response at selected locations does not exceed a specified limit.

It could therefore be expected that the responses and load levels around the notching ranges would exhibit similar characteristics like that of Figure 1(b). We investigated this on a typical example taken from a STM satellite structure (Structural Test Model) test campaign. The measured sinusoidal shaking table accelerations (input) for three different load levels are shown in Figure 2. The absolute response amplitudes of the 17 largest DOFs around resonance shown in Figure 3 confirm that the goal of constant output control was fulfilled satisfactorily. The corresponding transmissibilities obtained from dividing the responses by the table accelerations are shown in Figure 4. The similarity of these curves to the excitation force characteristics of the analytical example of Figure 1(b) is obvious. It can be observed that like in Figure 1(b) the transmissibilities do not exhibit significant distortions so that the utilization of classical linear modal extraction tools may be applied separately for each input level.

The technique described in the following utilizes this well established industrial test procedure to analyze the non-linearity based on controlled response at three different table acceleration levels all of them assumed to be constant within the notching frequency ranges.

## 2. THEORETICAL BACKGROUND

We describe in the following a technique aimed at predicting the responses to ***other than the measured*** input levels which is achieved by applying interpolation and extrapolation techniques to the modal data extracted from the transmissibilities measured at three measured input levels, i.e. the experimental data base is formed by ***three sets*** of natural frequencies, mode shapes, modal damping values and modal masses related to each load level.
At least three load levels have to be used for identifying equivalent linear modal data with classical EMA (Experimental Modal Analysis) techniques. This interpolation forms the basis of the ISSPA_NL code which is described in the following. The interpolation starts from classical linear modal analysis data extracted by any existing EMA code (we used our in-house ISSPA code, ref.[7]) from experimental frequency response functions (FRF) measured at three load levels.

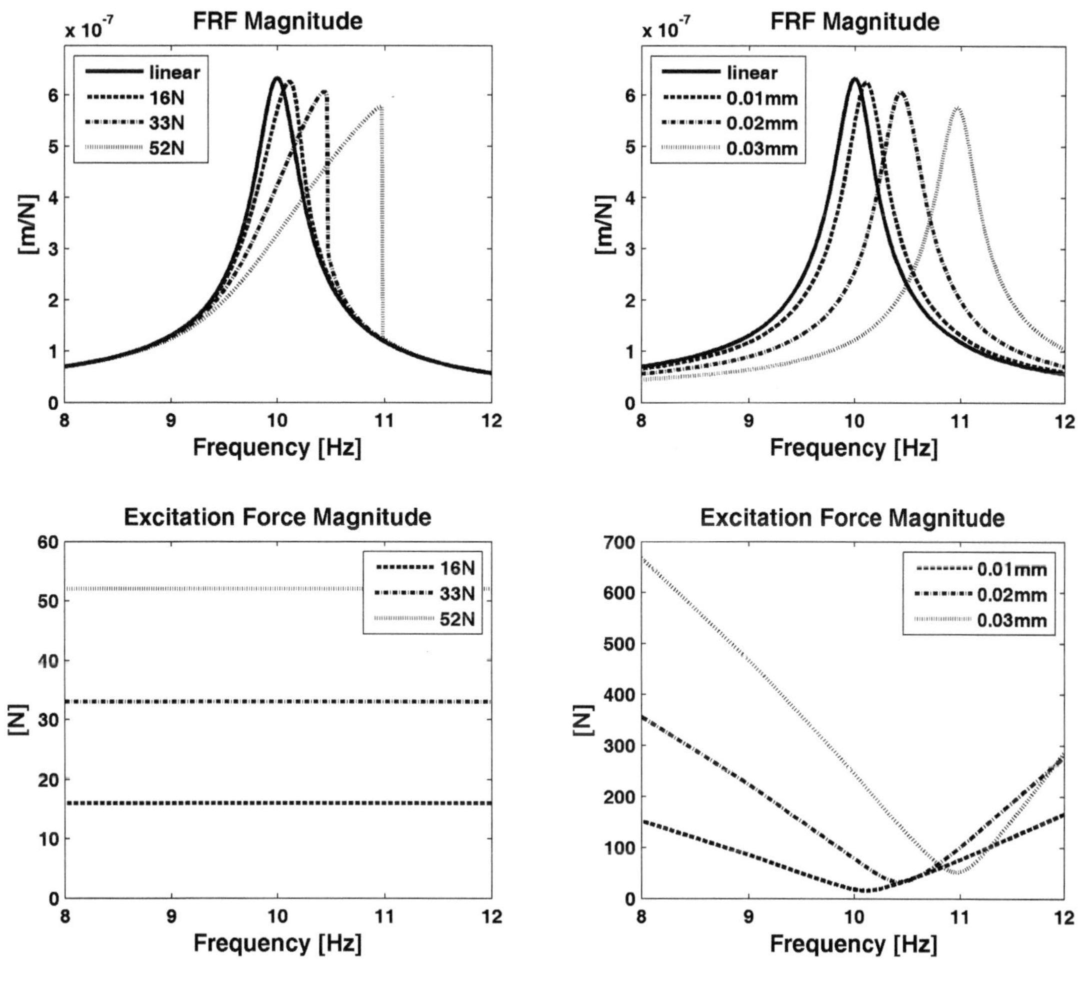


(a) constant excitation (b) constant response

Figure 1: FRFs obtained with constant excitation force levels (a) and constant response levels (b)

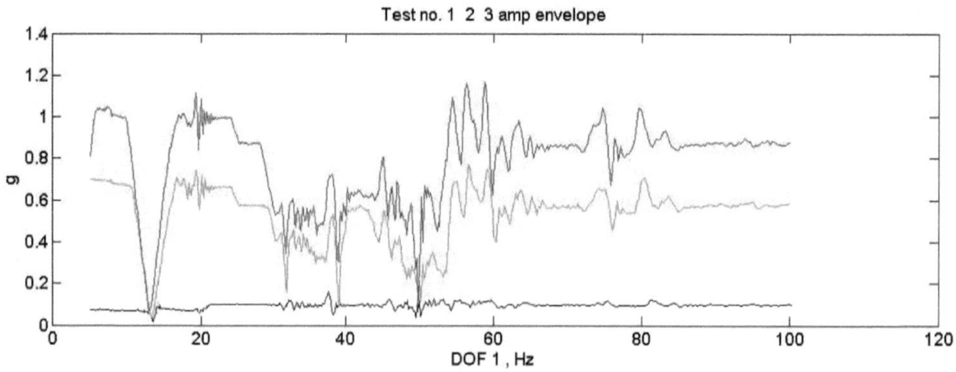


(a) Table acceleration (input) notched around resonances

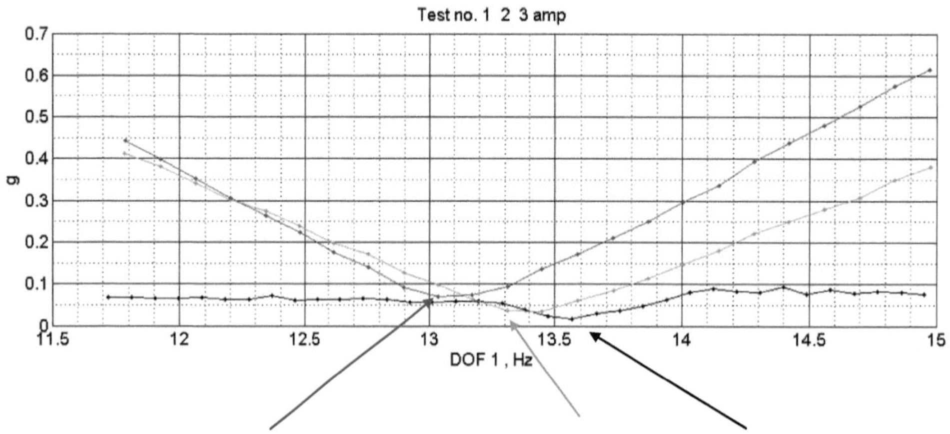


$f_{min}$= 13.10 Hz (qualification level), $f_{inter}$= 13.38 Hz (intermediate level), $f_{max}$= 13.56 Hz (low level)

(b) Table acceleration (pilot sensor) zoomed around first resonances

Figure 2: Sinusoidal shaking table accelerations (input) for low level (black), intermediate (green) and qualification level (red)

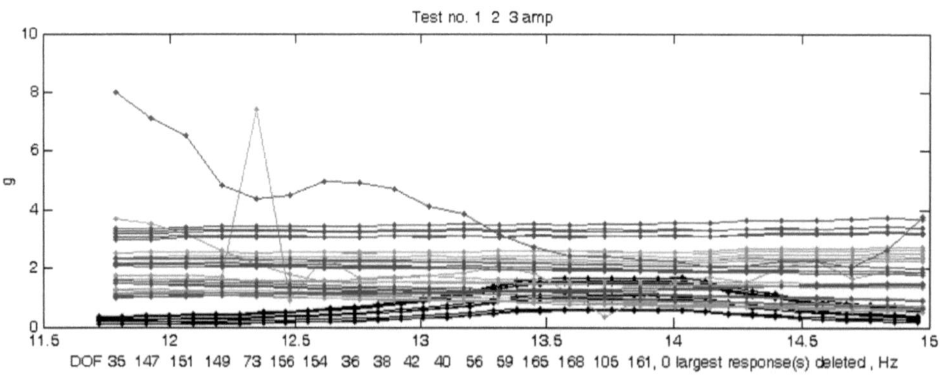


Figure 3: Nearly constant absolute response amplitudes of the 17 largest DOFs around resonance

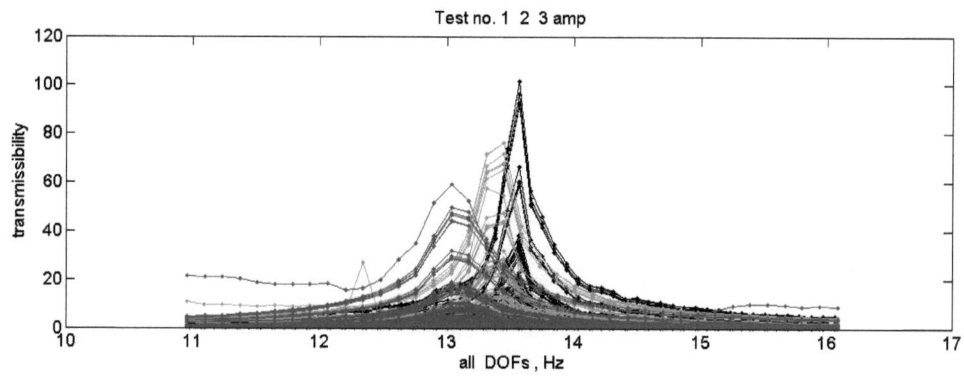


Figure 4: Amplitudes of the 17 largest transmissibilities
Low level (black), intermediate level (green), qualification level (red)

Figure 5 shows the interpolation and extrapolation scheme where the three data points for the variable *Z* on the vertical axis stand for any of the extracted natural frequencies, modal displacements, modal damping values or modal masses. The variable *A* stands for the load level used during the test (e.g. the g- level used on the shaking table). A distinction is made between interpolation levels $A_{int,i}$ (i=1,2...no. of interpolation levels) with the corresponding variable $Z_{int,i}$ located between levels $A_1$ and $A_3$ ($A_1 \leq A_{int,i} \leq A_3$ ) where a quadratic interpolation is performed whereas for levels $A_{low,i}$ and the corresponding variable $Z_{low,i}$ located below level $A_1$ and level $A_{up,i}$ and the corresponding variable $Z_{up,i}$ located beyond level $A_3$ where a linear extrapolation is performed based on the tangents in $A_1$ and $A_3$.

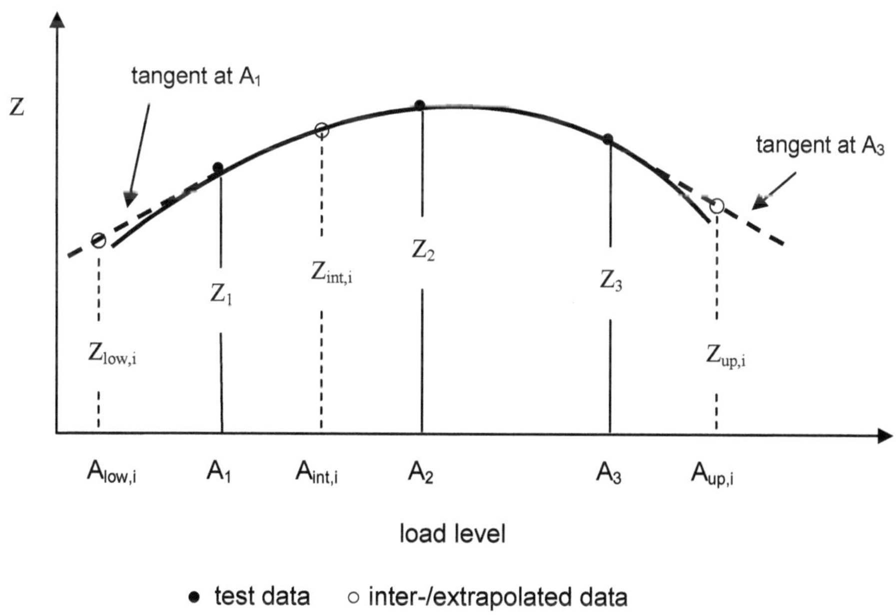


Figure 5. Interpolation scheme

It is clear that the extrapolation must be bounded by meaningful bounds. The validity range for the lower extrapolation is thus given by $A_{low} \leq A_{low,i} \leq A_1$ and $A_3 \leq A_{up,i} \leq A_{up}$ where the bounds $A_{low}$ and $A_{up}$ have to be specified by the user, e.g. $A_{low} = 0.9A_1$ and $A_{up} = 1.1A_3$.

The parabolic interpolation scheme between levels A1 and A3 is described by

$$Z_{int,i} = a_1 + a_2 A_{int,i} + a_3 A_{int,i}^2 = [A_{int}]_i^T \{a\} \tag{1}$$

Since this equation must also be valid at the measured load levels $A_1$-$A_3$ one gets three equations for determining the interpolation constants $a_1 - a_3$.

$$\begin{Bmatrix} Z_1 \\ Z_2 \\ Z_3 \end{Bmatrix} = \begin{bmatrix} 1 & A_1 & A_1^2 \\ 1 & A_2 & A_2^2 \\ 1 & A_3 & A_3^2 \end{bmatrix} \begin{Bmatrix} a_1 \\ a_2 \\ a_3 \end{Bmatrix} = [A]\{a\} => \tag{2}$$

$$\{a\} = [A]^{-1} \{Z\} \tag{3}$$

where the vector $\{Z\}$ holds the measured variables at the three load levels. Any interior variable $Z_{int,i}$ at any interior load level $A_{int,i}$ (i=1,2...n) is calculated from

$$\begin{Bmatrix} Z_{int,1} \\ \vdots \\ Z_{int,n} \end{Bmatrix} = \{Z\}_{int} = [A]_{int}^T [A]^{-1} \{Z\} \tag{4a}$$

where the matrix $[A]_{int}^T$ holds the n interior load levels in the form

$$[A_{int}]^T = \begin{bmatrix} 1 & A_{int,1} & A_{int,1}^2 \\ \vdots & \vdots & \vdots \\ 1 & A_{int,n} & A_{int,n}^2 \end{bmatrix} \tag{4b}$$

For the linear extrapolation scheme the tangents of the parabola at load levels $A_1$ and $A_3$ as shown in figure 5 are used which results in the interpolation formula for the lower and the upper range

$$Z_{low,i} = Z_1 + tan(Z(A_1))(A_{low,i} - A_1) \tag{5a}$$

$$Z_{up,i} = Z_3 + tan(Z(A_3))(A_{up,i} - A_3) \tag{5b}$$

with $tan(Z(A_1)) = a_2 + 2a_3A_1$ and $tan(Z(A_3)) = a_2 + 2a_3A_3$ where $a_2$ and $a_3$ denote the interpolation coefficients of eq.(3).

## 3. INDUSTRIAL APPLICATION

Test data were measured during a vibration test campaign of a typical STM satellite structure (Structural Test Model). The shaking table was driven in lateral direction at low, intermediate and qualification g – levels as shown in Figure 2. The data shown in Figure 4 represent the transmissibilities (transfer functions) calculated by division of the sensor responses by the pilot response which was chosen as the maximum of the four pilot sensors located on the shaking table.

The test data evaluation procedure consisted of the following steps:

(a) Analysis of the transmissibilities over the whole frequency range in order to identify the most significant response areas.

(b) Selection of a frequency range which is considered to be representative for non- linear behavior. We present here the frequency range around the fundamental resonance in lateral x- direction which exhibited the most significant peak shifts and large deviations of amplitudes.

(c) Identification of natural frequencies, mode shapes and modal damping values for each load level within the frequency range selected under (b) using ISSPA classical modal extraction technique

(d) Identification of load levels at resonance

(e) Application of ISSPA_nl for prediction of modal data at not measured input levels

**Identification of modal parameters**
The characteristics of the transmissibilities permitted the extraction the modal data separately for each load level using classical curve fitting procedure (ISSPA) which is based on the linearity assumption.

***Results***

| | | |
|---|---|---|
| Low Level | : fo= 13.56 Hz, | damping xsi = 0.8 % |
| Intermediate Level | : fo= 13.38 Hz, | xsi = 1.3 % |
| Qualification Level | : fo= 13.10 Hz, | xsi = 1.7 % |

The following MAC- values compare the mode shapes extracted from the three load levels:
MAC (low/ intermediate) = 93.2 %,
MAC (low/ qualification) = 93.0 %,
MAC (intermediate/qualification) = 96.9 %.

These numbers show

- a small variation of the resonance frequency
- a significant increase of the damping values with the load level
- a small influence of the load level on the mode shapes.

The correlation of test response and synthesised response was good (not shown here).

**Analysis of measured load levels**
The measured pilot sensor amplitudes for excitation in lateral x- direction for all three load levels are shown in Figure 2. From the zoom in Figure 2(b) It can be observed that the load levels are not constant in the notching frequency range around the first resonance (in particular for the intermediate and the qualification load levels). This plot shows the difficulty of applying the ISSPA_nl theory where it is assumed that the load levels are controlled to be constant within a given interpolation range. This interpolation range should cover the range between $f_{low} = 0.9f_{min}$ and $f_{up} = 1.1f_{max}$ where $f_{min}$ denotes the smallest resonance frequency and $f_{max}$ the largest resonance frequency identified from the three load levels which in the present case would mean a range between $f_{low}$= 11.8 Hz and $f_{up}$= 14.9 Hz. The plot in Figure 2(b) shows that this requirement is not fulfilled.

**Application of ISSPA_nl**

Although it was observed that the load levels are not constant in the interpolation frequency range an attempt was made for verifying the functionality of the ISSPA_nl code with the existing test data set. For this purpose it was assumed that the minimum load levels of the corresponding pilot responses measured at the resonance frequencies could be expanded to the whole interpolation frequency range. From Figure 2(b) these load levels were identified to be $L_{min}$= 0.0162 g, $I_{min}$ = 0.0350 g and $Q_{min}$ = 0.0715 g.

The application of ISSPA_nl now seeks to predict the modal data and the response at any other load level within $L_{min}$ and $Q_{min}$ and say +/- 10% outside of these load levels. Because of the non- constant load levels within the interpolation frequency range we restricted the prediction in the present application to two additional load levels within $L_{min}$ and $Q_{min}$, one at $X1_{add} = (L_{min} + I_{min})/2$ = 0.0256 g and the other at $X2_{add} = (I_{min} + Q_{min})/2$ = 0.0532 g. The basic modal data extracted by classical (linear) modal extraction technique using the transmissibilities of each load level are used as data points for the interpolation functions as described before.

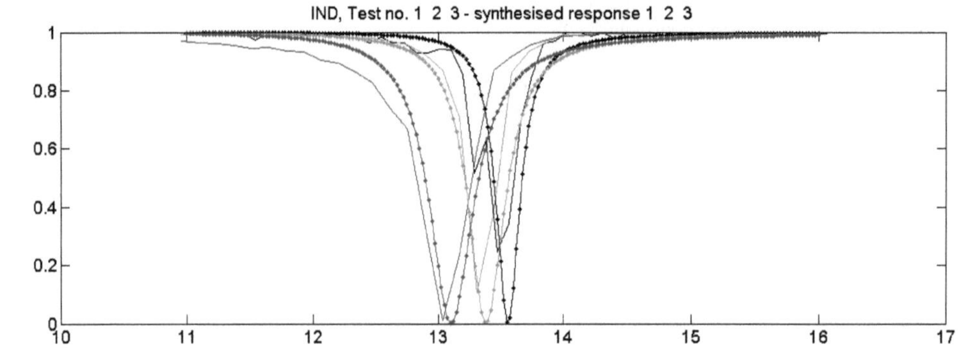


(a) Indicator functions

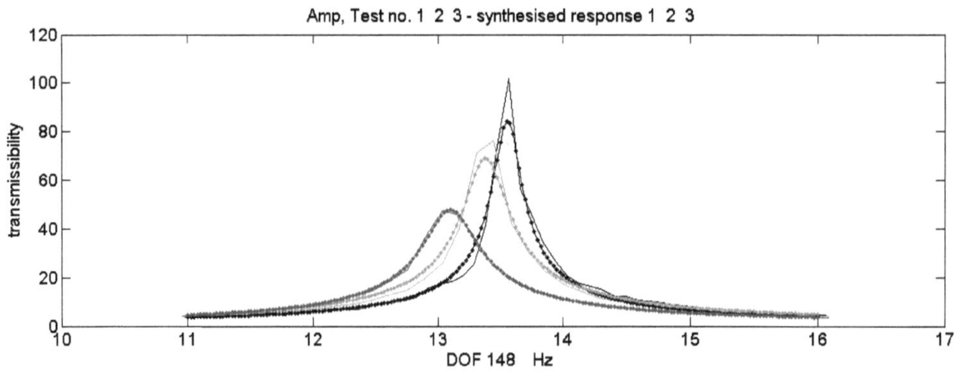


(b) Transmissibilities

Figure 6 Measured (——) and synthesised (• • •) indicator functions and transmissibility amplitudes low level (black), intermediate level (green), qualification level (red)

In Figure 6 we show the good correlation between test and modal synthesis for one representative measurement DOF on top of the structure (the correlation for other DOFs was similar) which confirms the assumption that due to controlled output testing the structure's response was linearized.

The results of interpolating the modal data by ISSPA_nl over the load levels between $L_{min}$= 0.0162 g and $Q_{min}$ = 0.0715 g are shown in Figure 7 (the modal displacements are not shown because of their low variability).

A considerable variation of the modal damping was observed whereas the maximum variation of the resonance frequency is restricted to not more than about 0.5 Hz. Using the modal data at the additional load levels $X1_{add}$ = 0.0256 g and $X2_{add}$ = 0.0532 g together with the modal data from the measured load levels yields the indicator functions and the transmissibilities shown in Figure 8.

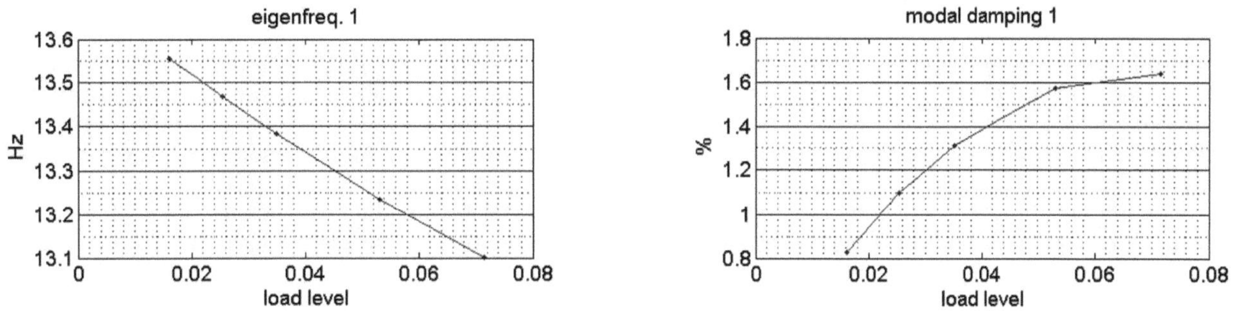


Figure 7 Resonance frequency and modal damping vs. load level [g]

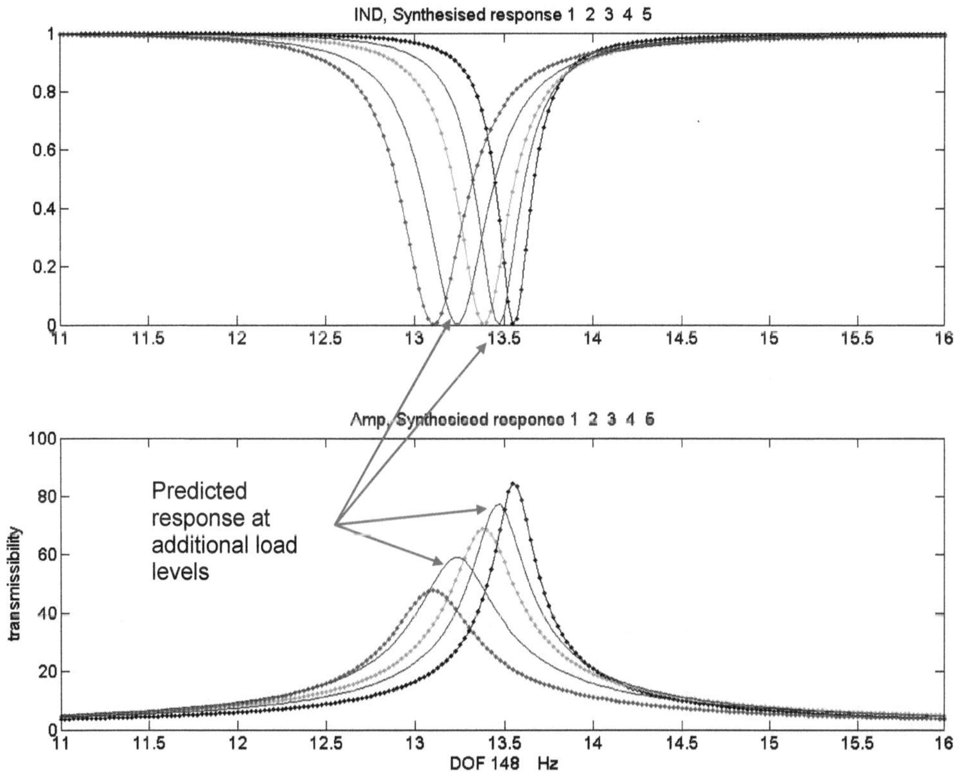


Figure 8 Indicator functions and transmissibilities using the modal data at the additional load levels (blue) together with the modal data from the measured load levels (qualification level red, intermediate level green, low level black)

It may be noted that the indicator functions and the transmissibilities at the additional load levels exhibit a physically meaningful interpolation between the measured levels and thus seems to confirm the achievement of the goals of ISSPA_nl. It should also be noted that the calculation of the load level dependent modal data was based on experimental modal data only, i.e. there was no mathematical model involved.

## CONCLUSIONS

Due to response control at selected sensor locations during a vibration test on a shaking table it can be observed that the transmissibilities do not exhibit significant non- linear distortions so that the utilization of classical linear modal extraction tools may be applied separately for each input level. However, the measured dynamic responses (transmissibilities) reveal peak shifts and amplitude changes depending on the notched input level of the base excitation which can be utilized to describe the non- linear behaviour. The technique described in the paper utilizes this well established industrial test procedure to analyze the non-linearity based on controlled response at three different table acceleration levels all of them assumed to be constant within the notching frequency ranges. The technique (implemented in the ISSPA_nl code) aimed at predicting the responses to ***other than the measured*** input levels which is achieved by applying interpolation and extrapolation techniques to the modal data extracted by classical (linear) experimental modal analysis from the transmissibilities measured at three measured input levels.

The procedure was applied using the test data from a STM satellite structure (Structural Test Model) test campaign. One characteristic frequency range was selected exhibiting non- linear behaviour around the dominant fundamental resonance. Subsequently, natural frequencies, mode shapes and modal damping values were extracted for each load level within the selected frequency ranges.

The most important goal of the study was to verify that controlled response vibration test data could be used for predicting the modal behaviour at other load levels than the measured levels. It was found that the predicted indicator functions and the transmissibilities at the additional (not measured) load levels exhibit a physically meaningful interpolation between the measured levels and thus seems to confirm the achievement of the goals. It could be of interest for future research to investigate how the load dependent modal data could be utilized for the parameter identification of a physical non- linear Finite Element model.

## REFERENCES

[1] K. Worden, G. Tomlinson: *"Nonlinearity in Structural Dynamics – Detection, Identification and Modelling"*, Institute of Physics Publishing, Bristol, 2001

[2] K. Vanhoenacker, J. Schoukens, J. Swevers, D. Vaes: *"Summary and Comparing Overview of Techniques for the Detection of Non-Linearities"*, Proc. of the International Conference on Noise and Vibration Engineering ISMA 2002, Leuven, Belgium, 2002

[3] J. Wong, J.E. Cooper, J.R. Wright: *"Detection and Quantification of Structural Non-Linearities*, Proc. of the International Conference on Noise and Vibration Engineering ISMA 2002, Leuven, Belgium, 2002

[4] Göge D., Sinapius M., Füllekrug U. and Link M.: *Detection and Description of Non-linear Phenomena in Experimental Modal Analysis via Linearity Plots.* Int. Journal of Non-linear Mechanics 40 (2005), 27- 48

[5] Ewins D.J.: Modal Testing: Theory, Practice and Application, $2^{nd}$ Ed.. Research Studies Press, Baldock,UK (2000)

[6] Böswald M. and Link M.: Identification of Non-linear Joint Parameters by Using Frequency Response Residuals. Int. Modal Analysis Conf. IMAC XXIII (2005)

[7] ISSPA User's Manual: http://www.uni-kassel.de/fb14/leichtbau/downloads/

Proceedings of the IMAC-XXVIII
February 1–4, 2010, Jacksonville, Florida USA


# Development of a Dynamic Model for Subsurface Damage in Sandwich Composite Materials

Ethan Brush, Douglas Adams
Purdue University, Center for Systems Integrity, Lafayette, IN 47905

## Nomenclature

| | |
|---|---|
| $k$ | Stiffness (m/N) |
| $m$ | Mass (kg) |
| $c$ | Damping factor (N/m/s) |
| $\omega$ | Frequency (Hz) |
| $\omega_n$ | Natural Frequency |
| *zeta* | Damping Ratio |
| $F$ | Force (N) |


## Abstract

Despite the advantages of lightweight composite materials over metals and metal alloys, many challenges remain in their inspection and failure prediction. Researchers have found that subsurface damage accumulation due to fiber breakage and delaminations can be detected by observing localized nonlinear vibrations in composite structures [1,2,3,4]. A mathematical model to describe these nonlinearities does not exist in the literature. This research presents an experimental approach to identify a single degree of freedom (SDOF) nonlinear dynamic system model that represents subsurface damages in lightweight sandwich composites as localized stiffness or damping nonlinearities.

The forced vibration of a damaged and pristine fiberglass laminated honeycomb sandwich composite panel is tested in this research. A test fixture is designed to clamp the panels in order to constrain the transverse oscillatory response of the material inside the clamped area to that of an equivalent SDOF system. Damage is introduced by cutting away the bond between the face sheet and core material. A linear vibration study is performed to quantify the changes in the mass, stiffness and damping properties of the material in the test fixture. A large decrease in the single degree of freedom stiffness is observed for the damaged case. Restoring force analysis of the damaged fiberglass material shows a clear bi-linear stiffness behavior.


## Section I: Introduction

Significant reductions in a composite material's strength can accompany subsurface damage that accumulates when fibers break, delaminations take place between layers, disbonds between face sheets and sandwich cores occur, or damage is experienced by the core. Damages like these are usually undetected in visual inspections and are cumbersome to detect with other non-destructive evaluation methods that require knowledge of where damage is located before applying localized inspection devices [5,6]. In addition, conventional methods of inspection such as ultrasonic and thermographic techniques equate damage to local changes in the geometry of the material. If there are local thickness and other variations due to manufacturing variability and structural design (e.g., stiffness laminate with stringers), these variations can interfere with the ability to detect damage. Therefore, wide area inspection methods that equate damage to changes in the mechanical properties of the material are desirable for localizing damage in composite structures. To enable these kinds of wide area inspections, there is a need for more accurate models that describe the effects damage has on the local mechanical properties of the material. Presently, researchers use assumptions about changes in the linear stiffness and/or damping as well as nonlinearities in stiffness and damping to describe material damage. The goal of this research is to develop an experimental methodology for determining forced mechanical vibration models to describe the effects of damage in composite sandwich structures. These models will aid in the application of future nondestructive evaluation techniques and structural health monitoring methods.

The material investigated in this research is a fiberglass sandwich composite with a plastic honeycomb core. This material is relatively inexpensive, yet representative of many different fiber reinforced sandwich materials. Subsurface damage is created by manually cutting the bond between the face sheet and core material on the edge of a specimen. This is shown in Figure 1. The sandwich panels are examined in pristine and damaged states in order to develop models for the two different scenarios.

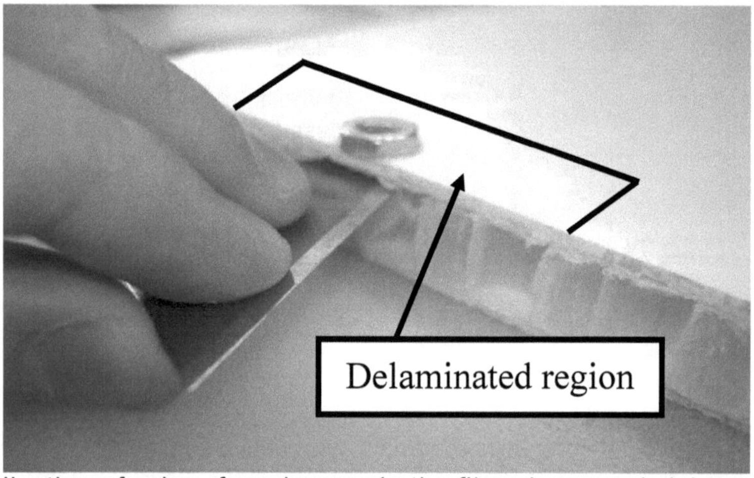


Figure 1: Application of subsurface damage in the fiberglass sandwich composite panel.

## Section II: Experimental Setup

A fixture was designed to constrain the test panel so that the response of the material behaved like a single degree of freedom system in the vicinity of the damaged area. Figure 2 shows the final fixture design. The lower plate was bolted to a large steel mass that acted to fix the boundary conditions so that the fixture did not contribute to vibration measurements.

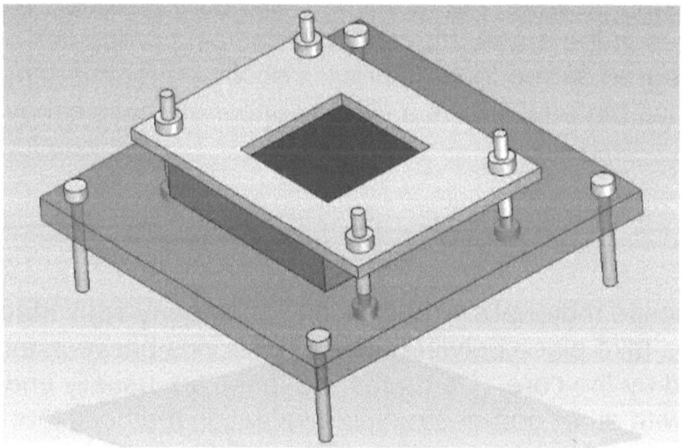

Figure 2: Fixture design for creating a SDOF system material response.

An impedance sensor was attached to the middle of the panel inside the fixture. The sandwich material was tested in two ways. First, frequency response function (FRF) measurements were made with the force input from a modal hammer and the acceleration response from the impedance sensor. The FRF amplitude measurements were then curve fit using a peak pick method to estimate single degree of freedom properties of the material. An illustration of the experimental model of the test setup is shown in Figure 3.

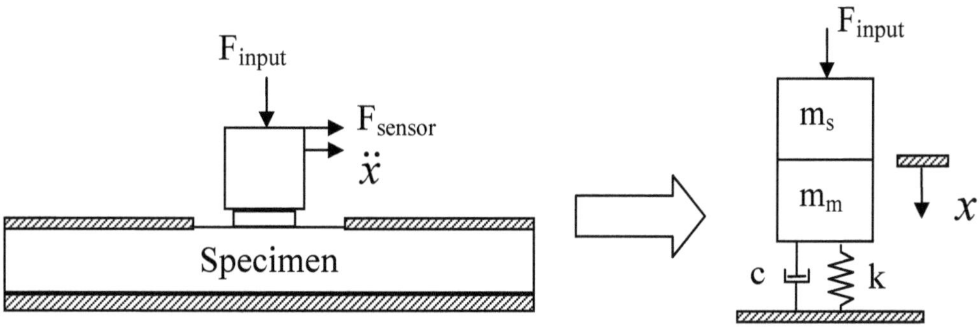


Figure 3: Experimental model of the material with the attached impedance sensor.

The second method in which the panel was tested employed the use of restoring force measurements in order to observe the relationship between force and displacement of the material. A small electro-dynamic shaker was used to excite the panel at single sinusoidal frequencies. A picture of the fixture and shaker setup for restoring force measurements is shown in Figure 4.

Figure 4: Test fixture and shaker for restoring force measurements.

Force and acceleration measurements from the impedance sensor were used to generate restoring force data. The acceleration response signal was integrated twice in order to obtain displacement data.

## Section III: Data Analysis

Frequency response function measurements were made in the test fixture on a healthy and damaged edge section of the fiberglass panel. A small modal hammer was used so that the excitation level was small enough to represent a nominally linear response of the material. A comparison of the FRFs for the two damage scenarios is shown in Figure 5.

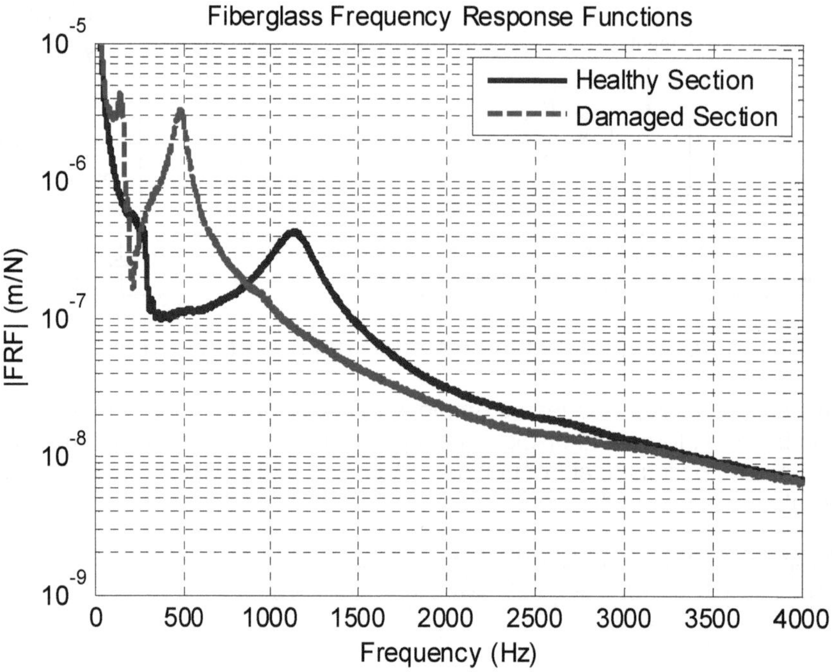


Figure 5: FRF measurements between the healthy and damaged sections of the panel.

A drop in natural frequency and increase in peak magnitude were observed in the FRF data. This suggested that the stiffness and damping were both decreasing. Data from Figure 5 was curve fit using a peak pick modal parameter estimation method. Frequency response functions for both sections indicated the magnitudes of the FRFs were increasing as the frequency neared 0 Hz. This phenomenon was attributed to clipping of the accelerometer sensor at very high frequencies due to the broadband excitation provided by the small modal hammer. The single degree of freedom nominally linear parameters of the healthy and damaged fiberglass composite are shown in Table 1.

Table 1: nominally linear parameters of each fiberglass sandwich test specimen.

| | k (N/m) | m (kg) | c (Ns/m) | $\omega_n$ (Hz) | zeta |
|---|---|---|---|---|---|
| Healthy | 1.34E+06 | 0.0264 | 31.84 | 1136 | 0.085 |
| Damaged (mode 2) | 2.57E+05 | 0.0277 | 9.8 | 485 | 0.058 |
| Damaged (mode 1) | 9.69E+04 | 0.1181 | 25.28 | 144 | 0.118 |

It was observed in Figure 5 that the frequency response functions measured above the damaged section contained two peaks. It is believed that this result can be attributed to two different stiffness values present in the material. Both peaks were examined for SDOF parameter estimates as shown in Table 1. It is clear from Table 1 that the SDOF stiffness and damping parameters of the fiberglass panel decreased significantly. The SDOF mass parameter did not change significantly between the 485 Hz peak in the damaged section and the 1136 Hz peak in the healthy section of the panel.

Restoring force data was collected for the healthy and damaged sections of the test panel with a sinusoidal input of 100 Hz from the electro-dynamic shaker. The restoring force response of the panel was measured at 5 different excitation amplitudes. The results for the healthy section of the panel indicate that the force and displacement relationship is largely linear for all excitation amplitudes as illustrated in Figure 6.

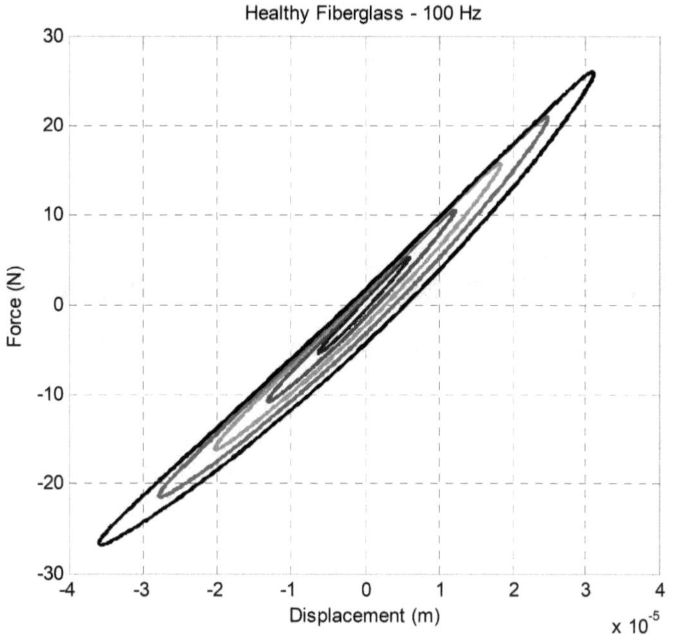


Figure 6: Healthy fiberglass restoring force curves at 100 Hz.

Restoring force data for the damaged section of the fiberglass panel is shown in Figure 7. The restoring force of the damaged fiberglass clearly showed a nonlinear behavior as the excitation amplitude was increased. The direction of positive displacement was towards the fiberglass material. Figure 7 indicated that the stiffness when forcing into the material was more than the stiffness when the excitation pulled away from the material. It could be seen that for the same displacement the force amplitude was almost double in the positive direction. The change in apparent stiffness values occurred at displacements greater than zero. This suggested that the larger stiffness value was due to the face sheet coming in contact with the honeycomb core. It was believed that the gap created by the disbond between the face sheet and core explained why the transition in stiffness values occurred at a displacement greater than zero.

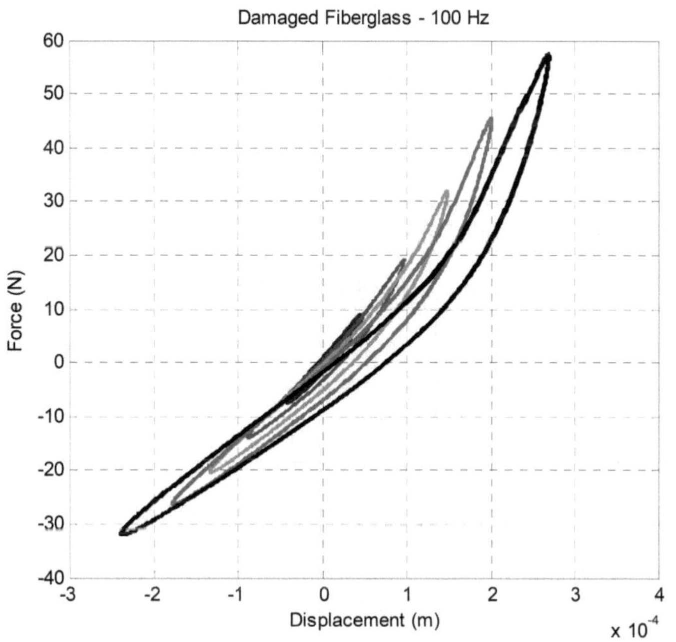


Figure 7: Damaged fiberglass restoring force curves at 100 Hz.

Observations of the restoring force behavior exhibited by the damaged fiberglass material at 100 Hz suggested that a bi-linear stiffness model was appropriate for describing the experimental data. Figure 8 illustrates the proposed system when the force acts to open and close the disbond region.

Figure 8: Proposed opening and closing of the disbond between the face sheet and core of the panel

When the force acts to open the disbond region the resistance to deflection comes from the bending stiffness of the top face sheet. When the disbond gap is forced closed, there are two stiffness values present. The first stiffness is contributed by the bending of the face sheet. The second stiffness is contributed by the honeycomb core material. These two springs act in parallel with one another when the disbond gap is closed.

Figure 9 shows the proposed bi-linear stiffness model of the fiberglass sandwich composite. Because the two springs are in parallel, the equivalent stiffness of the material when forcing the gap closed will be the summation of the two individual stiffness elements.

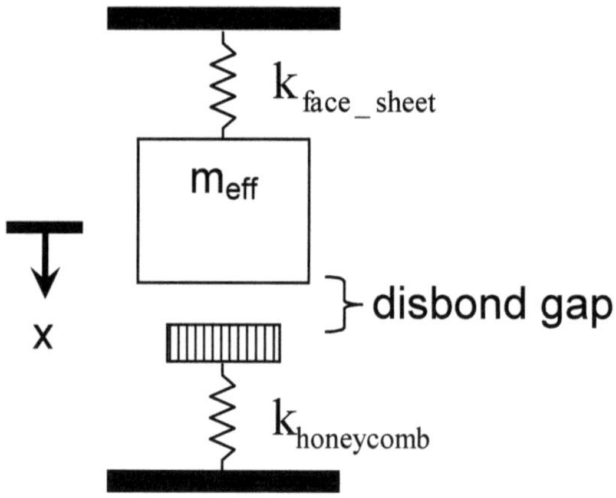


Figure 9: Proposed bi-linear stiffness model of the fiberglass disbond.

A Simulink model was developed in order to simulate the forced response of the system in Figure 9. Parameter values chosen for the simulation are shown in Table 2.

Table 2: Bi-linear model simulation parameters.

| $k_1$ (N/m) | $k_2$ (N/m) | m (kg) | c (Ns/m) | Disbond gap (m) |
|---|---|---|---|---|
| 8.68E+04 | 8.52E+05 | 0.251 | 50 | 1.0E-04 |

The two effective stiffness parameters were obtained by estimating the slope stiffness values from the experimental data in Figure 7. The disbond gap was chosen to represent the point at which the experimental data transitioned to a higher slope. The mass and damping coefficients were chosen in an ad hoc manner such that the simulation compared favorably with the experimental restoring force data. The result of the simulation for an input of 30 N at 100 Hz is shown in Figure 10.

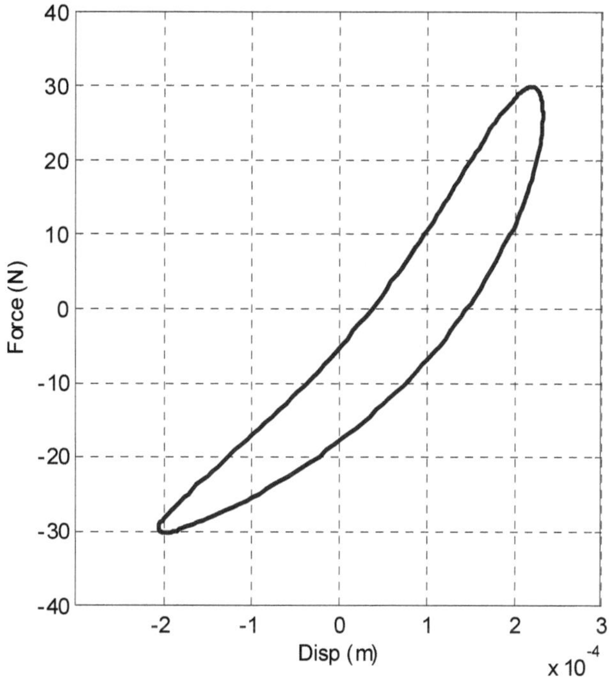


Figure 10: Bi-linear stiffness model simulation results.

It was found that the simulated restoring force diagram compared reasonably well with the experimental data. Therefore, it was concluded that a bi-linear stiffness model was appropriate for describing the effects of subsurface damage in fiberglass composite sandwich material

## Section IV: Conclusions

A fiberglass sandwich composite panel was subjected to subsurface damage by cutting the bond between the face sheet and honeycomb core material. A test fixture was designed in order to constrain the response of the sandwich panels to that of an equivalent single degree of freedom system. A study was performed to quantify the nominally linear changes in the mass, damping, and stiffness properties of the material in the test fixture. Significant reductions in the stiffness and damping coefficients were estimated after disbonding was created in the fiberglass material. The estimated change in mass was relatively small.

The forced vibration of the damaged material was then investigated using restoring force analysis. Experimental restoring force data with the fiberglass sandwich composite material lead to the implementation of a bi-linear stiffness model for describing the observed dynamic responses. Now that a nonlinear model has been proposed for this type of damaged material, further research should consider the implementation of this model with existing structural health monitoring methods that rely on the presence of nonlinear dynamics to detect damage. Also, this experimental method can be improved with the use of robust parameter estimation algorithms to more accurately estimate SDOF model coefficients.

## Section V: References

[1] Underwood, S., Adams, D., Koester, D., Plumlee, M., Zwink, B., "Structural Damage Detection in a Sandwich Honeycomb Composite Rotor Blade Material Using Three-Dimensional Laser Velocity Measurements." Proceedings of the American Helicopter Society 65th Annual Forum, May 27-29, 2009.

[2] Yoder, N.C., Adams, D.E., Triplett, M., "Multi-Dimensional Sensing for Impact Load and Damage Evaluation in a Carbon Filament Wound Canister", Materials Evaluations, 66(7), 756-763, 2008.

[3] Zumpano, G., Meo, M., “Damage Detection in an Aircraft Foam Sandwich Panel Using Nonlinear Elastic Wave Spectroscopy”, Composites and Structures, 86, 483-490, 2007.

[4] Zwink, B.R., Adams, D.E., Evans, R.D., Koester, D.J., “Wide-Area Damage Detection in Military Composite Helicopter Structures Using Vibration - Based Reciprocity Measurements”, Proceedings of the 27th International Modal Analysis Conference, 298-309, 2009.

[5] Kerschen, G., Worden, K., Vakakis, A.F., Golinval, J.C., “Past, Present and Future of Nonlinear System Identification in Structural Dynamics”. Mechanical Systems and Signal Processing, 20, 505-592, 2006.

[6] Haroon, M., Adams, D.E., “Time and Frequency Domain Nonlinear System Characterization for Mechanical Fault Identification”, Journal of Nonlinear Dynamics, 50, 387-408, 2007.

Proceedings of the IMAC-XXVIII
February 1–4, 2010, Jacksonville, Florida USA


# Transmissibility Analysis for State Awareness in High Bandwidth Structures Under Broadband Loading Conditions

Douglas Adams[1], Nathanael Yoder[1], Charles Butner[1], Richard Bono[2], Jason Foley[3] and Janet Wolfson[4]

## Nomenclature

| | |
|---|---|
| $z$ | Axial measurement degree of freedom |
| $r$ | Radial measurement degree of freedom |
| @ | Input degree of freedom |
| # | Output degree of freedom |
| $K_p$ | Stiffness parameters in lumped parameter model |
| $M_p$ | Mass parameters in lumped parameter model |
| $\omega$ | Circular frequency in rad/s |
| $H_{pq}(\omega)$ | Frequency response function for input $q$ and output $p$ |
| $X_p(\omega)$ | Response displacement for degree of freedom $p$ |
| $F_q(\omega)$ | Forcing function for degree of freedom $q$ |
| $T_{p1,p2(q)}(\omega)$ | Transmissibility function between $p2$ and $p1$ for input $q$ |

## Abstract

Under broadband loading, structures that are comprised of multiple stiff sub-components with interfaces exhibit complex dynamic responses. These responses can affect both the reliability and functionality of the structure. For example, if sensors within the structure are utilized to monitor its operational response for the purpose of estimating the loading state, it is essential to understand the relationship between those sensor signals and the dynamic complexity across the assembly interfaces. This paper describes experiments that were performed on a massive and stiff structural housing containing lower mass sub-components, which are assembled through lower stiffness, nonlinear threaded interfaces. It is shown that the lower frequency range of the multi-component specimen exhibits primarily rigid body motion and synchronous motions of the internal canister oscillating on a threaded interface. The mid-frequency range exhibits complex asynchronous motions as these various internal components oscillate out of phase with one another. Transmissibility functions, which are the ratios of measured frequency response functions, are used to analyze the transfer path through the structural housing and sub-components. It is demonstrated that under various preloads, the contribution of each interface to the overall transmission of load/motion varies. These variations in transmission due to differences in preload are shown to be of the same order as the softening nonlinear variations due to a change in the excitation force amplitude.

## Section I: Introduction

Threaded joints are pervasive in aerospace structures including munitions, which generally contain at a minimum a housing and internal instrumentation assembly that interact through a threaded interface of some kind. The instrumentation assembly usually contains sensing electronics that must be designed to withstand and observe broadband loads experienced by the housing. When the preload across these threaded joints is modified, the structural dynamics of the munitions can vary significantly leading to differences in how the sensors respond to

[1] Purdue University, Center for Systems Integrity, Lafayette, IN 47905
[2] The Modal Shop, Inc., Cincinnati, OH, 45241
[3] Air Force Research Laboratory, Munitions Directorate, Fuzes Branch, Eglin Air Force Base
[4] Applied Research Associates, Niceville, FL, 32758

the operational environment. In addition, variations in the amplitude, bandwidth, and distribution of excitation forces that act upon the housing lead to significant variations in the munition's structural dynamic response.
In this paper, these variations in the structural dynamic response are experimentally examined for the first time in the literature using transmissibility response functions. Transmissibility functions are calculated by taking the ratio of frequency response functions, which are measured between driving point and cross point degrees of freedom. Transmissibility analysis is useful to consider in bodies that are connected across threaded interfaces because transmissibility functions highlight internal resonances of groups of components within the structure as a whole. These functions also indicate where forces (and motions) are amplified and attenuated in passing from the driving point to the cross point degree of freedom making them ideal for use in studying the reliability and observability that is afforded by certain sensor package designs in munitions. First, transmissibility functions are calculated and analyzed in structural dynamic experiments in which variations in the preload across a series of threaded interfaces and amplitudes and locations of excitation forces are considered. Then a simplified lumped parameter model of a two degree of freedom structural dynamic system is analyzed to help interpret the results of the experiments.

A review article by Ibrahim and Pettit (2005) defined many of the issues associated with bolted joints in structural dynamics including the uncertainty and non-smooth nonlinear characteristics that are exhibited by such joints. This review article described the energy dissipation mechanisms across bolted joints as well as the linear and nonlinear identification of these joints. In the work presented here, experimentally measured transmissibility functions are used to characterize the filtering properties of threaded interfaces and variations in the dynamic response due to changes in both joint preloads and excitation amplitude. The work by Doebling et al. (2002) is also relevant to the study presented here because they considered the development of a transient dynamic model of a set of stiff components that interact through a threaded interface. The work by Hess and Sudhrikashyap (1996) is also relevant because they too considered the axial dynamic response of structures that interact through threaded fasteners as in the test specimen that is examined here. Finally, the work by Rashquinha and Hess (1997) on the modeling of nonlinear dynamic interactions across threaded joints is related to the work presented in this paper because they considered the design of joints to prevent loosening.

## Section II: Experimental Setup

Figure 1 shows a schematic of the top view (left) and side view (right) of the specimen that was tested in this paper. The cylindrical housing is made of steel and possesses a conical nose with a protruding tip. Five uniaxial 10 mV/g PCB 353B17 accelerometers (10 kHz) were mounted on the tip and nose normal to those surfaces. A triaxial 100 mV/g PCB 356A32 accelerometer (5 kHz) was mounted to the base of the housing. Modal impacts were applied at all points denoted with a "@" symbol using a 60 mV/lb PCB 086C01 modally tuned hammer and a PI piezoceramic stack actuator driven with a voltage source. The "z" and "r" notation is used to denote the direction of applied input forces at locations on the housing where multiple degrees of freedom could be measured. These denote modal impacts in the axial and radial directions, respectively. The serial numbers of the sensors are also indicated in Figure 1. The specimen was supported as shown in Figure 2(a) and (b) using bungee cords while loose nylon straps were installed in the event the bungee cords broke. Figure 2(c) shows the modal impact and piezoceramic actuator that were applied to the tip of the housing along with a single axis accelerometer to measure the driving point frequency response function. Figure 3 shows the internally installed sensors on the web of the canister (channels 10-12, 13, and 16-17), retainer ring (channels 14 and 18), and lock ring (channel 15). Two back to back triaxial accelerometers (PCB modal and Endevco shock models) were placed on the web for comparison (Figures 4(b,c)). Only two of the shock accelerometers (z and x directions) were powered. In addition, 60 kHz PCB 352A60 (10 mV/g) accelerometers were placed on each side of the interfaces on components within the housing (Figure 4(a)). An Agilent VXI data acquisition system with Matlab interface was used to acquire the data through two 1433 cards, each with 8 channels sampling at 30.72 kHz and 3 averages. The modal hammer and PCB accelerometers were connected through ICP channels, and the piezoceramic actuator voltage and shock accelerometer were connected through voltage channels (the triaxial shock accelerometer channels were also connected to a DC power supply).

Modal impact tests were performed both with the lock ring tightened to 250 ft-lb torque and hand tightened to characterize the frequency response out to approximately 6000 Hz. In these modal impact tests, 18 locations on the housing were impacted normal to the surface, and impacts were applied in both the radial and axial directions at the base of the housing. Tip impacts were also applied in the axial direction at the tip in the 0, 90, 180, and 270 degree locations to examine the differences in the coupling to off-axis responses due to the slightly off-axial z impacts in these locations. Tests using the piezoceramic stack actuator were also performed to characterize the higher frequency response beyond 6000 Hz out to 18 kHz. At the 250 ft-lb torque preload condition, four different impact levels were also applied to study the degree and nature of the nonlinear behavior due to the interfaces.

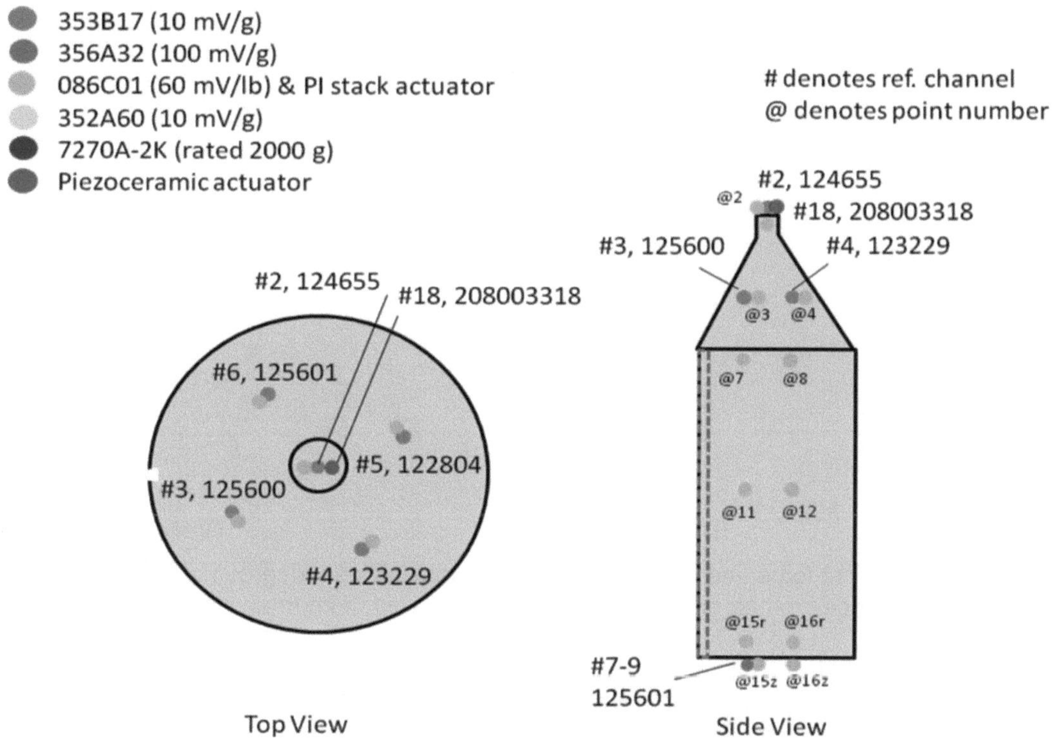

**Figure 1:** Top view and side views of cylindrical test specimen housing indicating force input and acceleration sensor output measurement degrees of freedom.

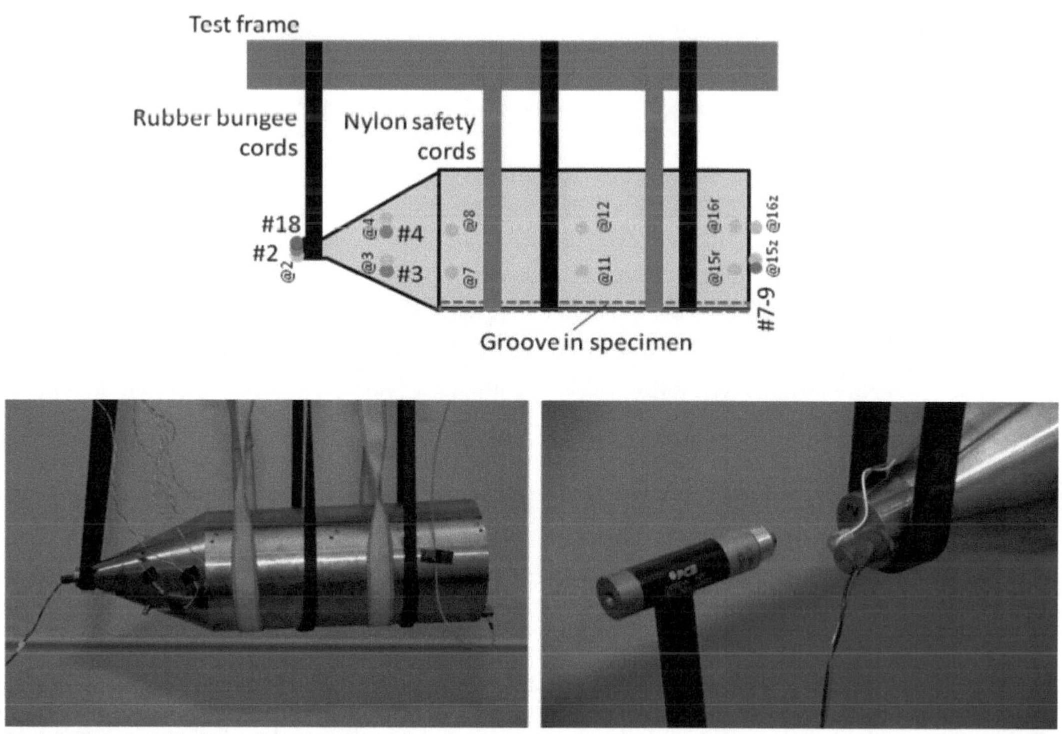

**Figure 2:** (a) Schematic and (b) photograph of specimen supported by bungee cords, and (c) modal impact hammer and piezoceramic stack actuator used to apply nearly axial input forces on tip of specimen.

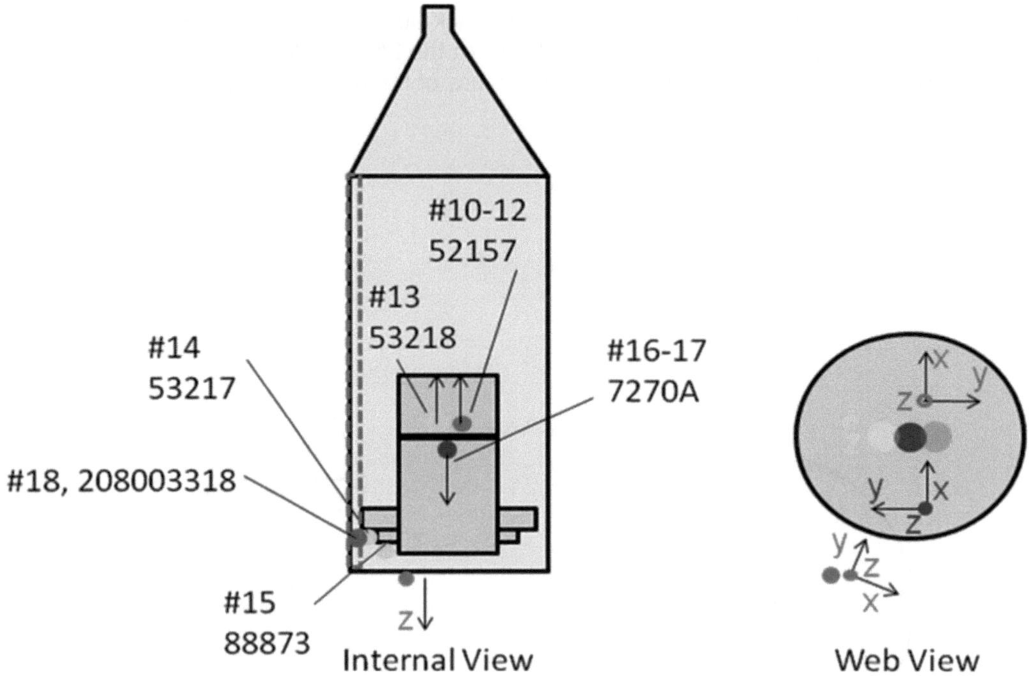


**Figure 3:** Internal view of input force and output acceleration measurement degrees of freedom showing two triaxial accelerometers placed back to back on web of internal canister.

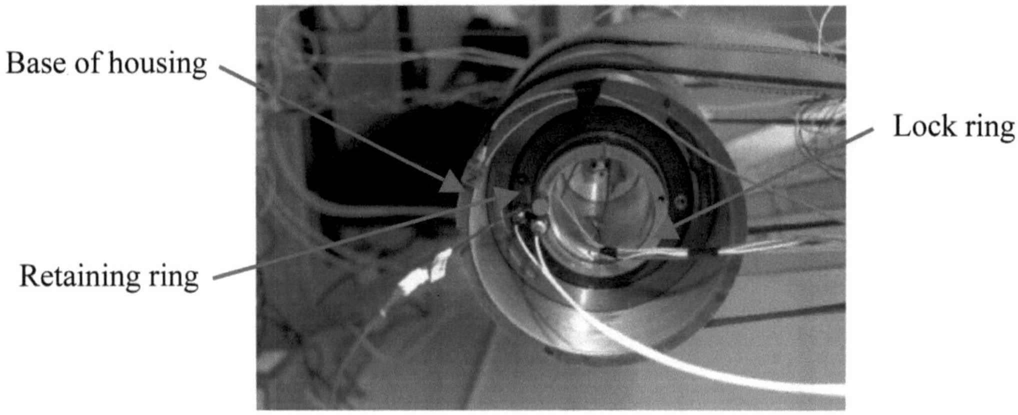

**Figure 4:** Photographs corresponding to Figure 3 showing (a) 60 kHz uniaxial accelerometers glued to retainer ring and lock ring along with stack actuator, (b) shock rated triaxial accelerometer stud mounted on base of web, and (c) modal triaxial accelerometer glued to top of web along with uniaxial 60 kHz accelerometer.

## Section III: Data Analysis

Figure 5 shows the average input autopower spectra for five different input degrees of freedom 2 (tip) and 3, 4, 5, and 6 (nose). The 20 dB roll off frequency for each input force measurement was approximately 6000 Hz; therefore, the frequency range from 0-6000 Hz in the modal impact data was the focus of this analysis.

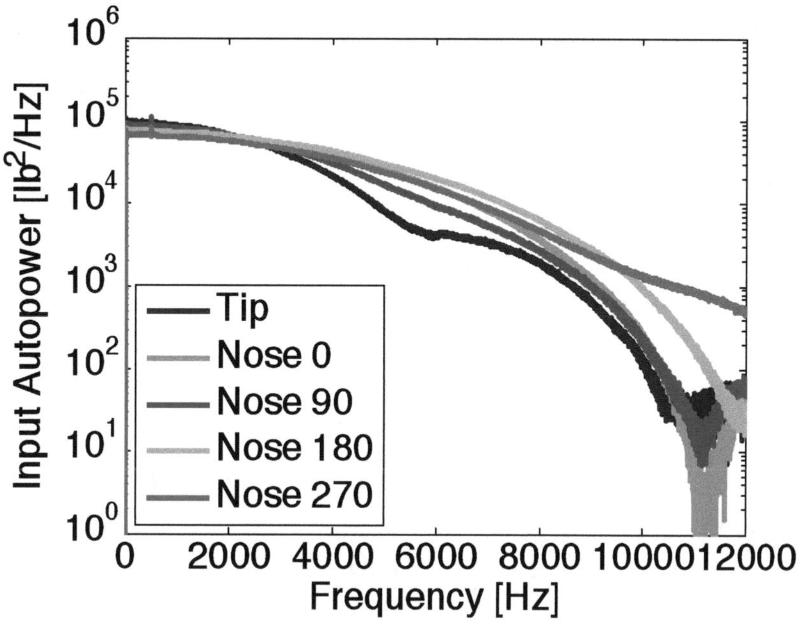


**Figure 5**: Input autopowers of modal impact forces on tip and at four locations around nose.

Figure 6 shows the frequency response function magnitudes between an axial impact force applied at the tip and acceleration responses measured at the tip ( ▬ ), the base of the housing ( ▬ ), and the web ( ▬ ) for the hand tightened lock ring condition. Note that the magnitudes of all the frequency response functions approach 0.01 g/lb. Since acceleration/force frequency response functions for an unrestrained body approach 1/m, where m is in units of lbs, this low-frequency value translates into an approximately 100 lb mass of the specimen, which is consistent with the actual value. The fact that all three of these frequency response functions approach the same low-frequency value suggest that the housing and its contents move as a rigid body at low frequency by swinging on the supports shown in Figure 2(b). The frequency response magnitude at the tip approaches an anti-resonance near 1400 Hz, whereas the corresponding magnitude for the web begins to approach the specimen resonance at 2210 Hz. Moreover, the anti-resonance at 1400 Hz appears to be the result of a superposition of the rigid body mode near 0 Hz and the flexible body mode at 2210 Hz.

Based on an analysis of the quadrature (imaginary) portion of the frequency response function at the 2210 Hz resonant frequency (see Figure 7), the mode shape is found to involve strong synchronous longitudinal motion of the web, lock ring, and retaining ring and much less motion of the tip and housing base. There are additional anti-resonances in the driving point tip frequency response function near 1470 Hz, 1600 Hz, and 1950 Hz. At all of these resonant frequencies, the web frequency response magnitudes again indicate that the web moves with significantly larger amplitude (in phase with the lock ring and retaining ring) than the base of the housing or the tip similar to the case for the resonance at 2210 Hz. Because the primary interest in this paper is on broadband response to impulsive type forces, it can be surmised that the response from 1400 to 2210 Hz is governed by canister motion with strong motion across the interface between the housing and retaining ring (see Figure 4(a)). It is anticipated that nonlinearities will be present in the modal response at 2210 Hz, and even in the lower resonant frequencies at 1600 Hz, due to this interfacial motion between the housing and retaining ring.

In the frequency range from 2200-6000 Hz, the web responds asynchronously with the lock ring and retaining ring. For example, at 2790 Hz, the web moves out of phase with the lock ring with significantly more amplitude than the base of the housing or the tip. Similarly, at 3200 Hz, there is large relative motion between the web, lock ring, and retaining ring. At 3500 Hz, the tip moves with considerably more amplitude than all of the other components. At 4750 Hz, the web resonates at a larger amplitude than all of components. This result suggests that in this frequency range from 2200-6000 Hz, there is considerable motion across the threaded interfaces. Above 6000 Hz, the relative motions across the interfaces are even more significant. It is anticipated that nonlinearities will be present in the modal responses from 2200-6000 Hz due to this extensive amount of interfacial motion.

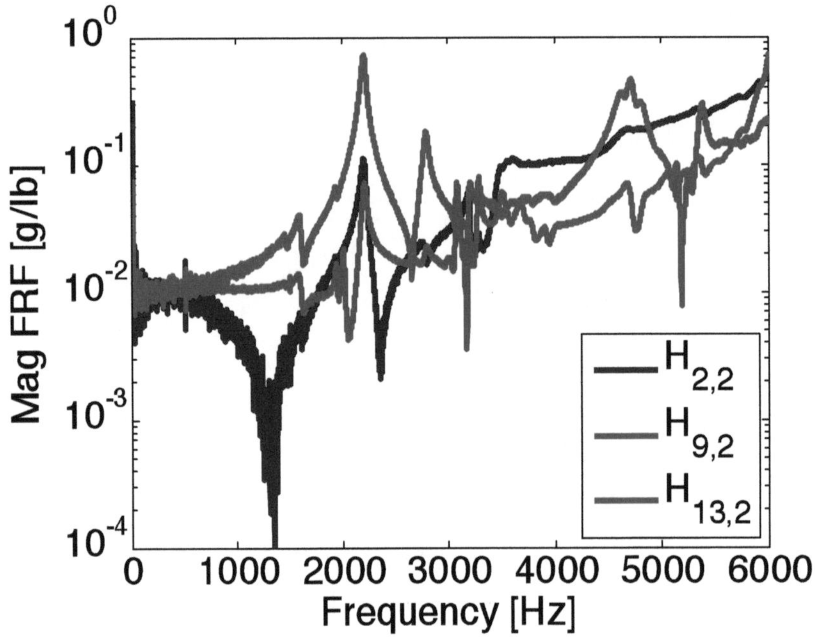


**Figure 6:** Frequency response functions between axial impact force at tip and tip (2), base of housing (9), and web (13) showing differences in dynamic response due to location.

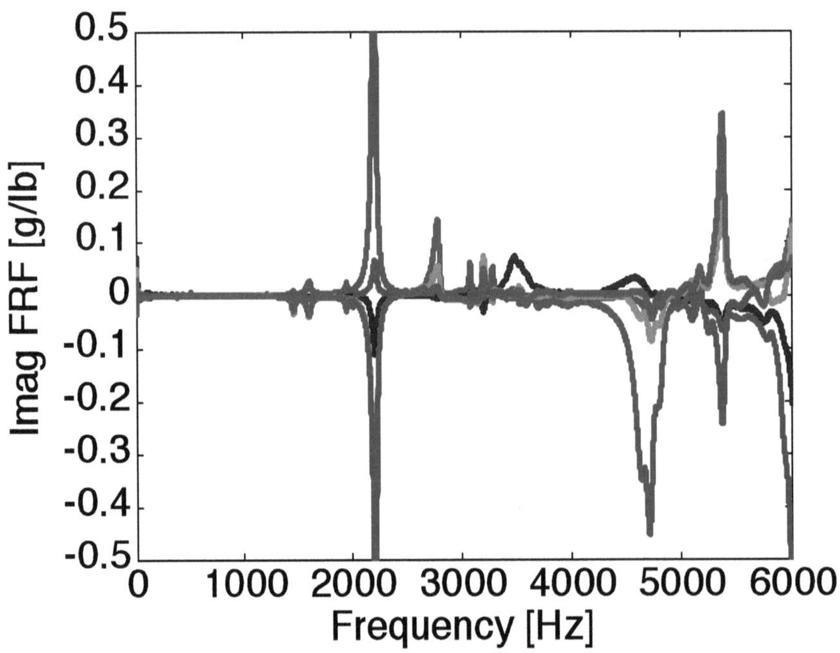


**Figure 7:** Imaginary part of frequency response function between axial impact force at tip and tip, base of housing, web, retainer ring, and lock ring showing synchronous motion of canister (hand tight condition).

When the quadrature components of the frequency response functions with 250 ft-lb tightened lock ring are considered (Figure 8), similarities with the hand tightened modal deflection shapes, which were discussed above, can be identified. Below 2210 Hz, the two sets of frequency response functions are quite similar. The modal deflection shape at 1470 Hz exhibits larger motions on the lock ring ( ▬ ) and retaining ring ( ▬ ) than on the web in the hand tightened case. In the frequency range from 2210 to 4200 Hz, the modal deflection shapes are quite different. For example, at 2790 Hz, the lock ring rather than the web moves with the largest modal deflection. Note that the tip frequency response ( ▬ ) is nearly the same before and after the lock ring is tightened to 250 ft-lb as is the housing base frequency response ( ▬ ) as expected since the housing was not modified when tightening the lock ring. In contrast, the web, lock ring, and retaining ring frequency responses between 2210 and 6000 Hz change significantly due to the change in preload across the threaded interface. This result again suggests that nonlinearities will be dominant in the frequency range above 1470 Hz.

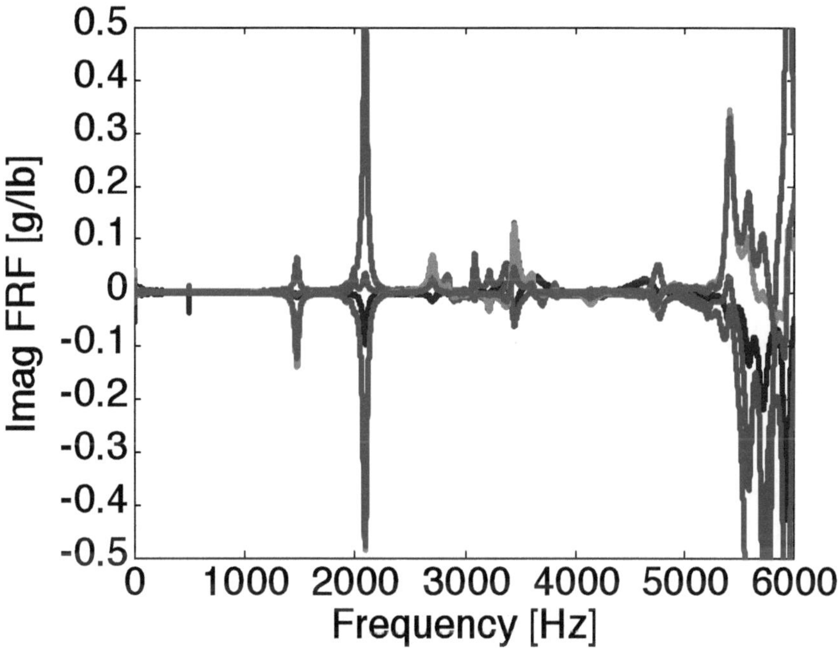


**Figure 8:** Imaginary part of frequency response function between axial impact force at tip and tip, base of housing, web, retainer ring, and lock ring showing synchronous motion of canister (250 ft-lb tight condition).

Another way to view the forced response is to calculate the ratios of frequency response functions between the input force degree of freedom and output response degree freedom. Figure 9(a) shows four pathways from the tip to various response measurements on the web, retaining ring, lock ring, and housing base. For each of these pathways, the transmissibility functions were calculated using the measured frequency response functions. Figure 9(b) shows a comparison of these transmissibility functions in the hand tightened condition. Below 1400 Hz, the transmissibility functions nearly lie on top of one another indicating that the transmission along each path is the same because the specimen response is dominated by a rigid body motion as the specimen swings on the bungee cords. This result is consistent with the frequency response function analysis discussed above. The peak near 1400 Hz indicates a frequency at which the transmission of motion from the tip to the other components is greatly amplified. In other words, the tip is nearly behaving dynamically like a fixed point at this frequency and the other components oscillate about his fixed point. Between 1400 and 2500 Hz, the web, retaining ring, and lock ring move together as discovered previously using the quadrature component of the frequency response functions. As these components move together, relative motion takes place across the interface between the specimen housing ( ▬ ) and the retaining ring ( ▬ ). In the frequency range from 3500 to 5000 Hz, the web ( ▬ ) responds with the largest deflection amplitude. Likewise, at 2800 Hz, the web responds with the largest amplitude relative to the tip of the specimen. The transmissibility plots reveal that, starting at around 2800 Hz, the relative motions across the interfaces between the web and lock ring and between the lock ring and retaining ring increase substantially. These results are consistent with the observations made using the frequency response functions.

By plotting the transmissibility functions for each of the pathways that are shown in Figure 9(a), the relative contributions of each of the interfaces can also be easily identified. For example, the frequency range from 3500 to 5000 Hz reveals that the largest relative motions are between the housing and retaining ring and between the lock ring and the web – by comparison, the relative motion between the lock ring and the retaining ring is lower in amplitude than these other motions.

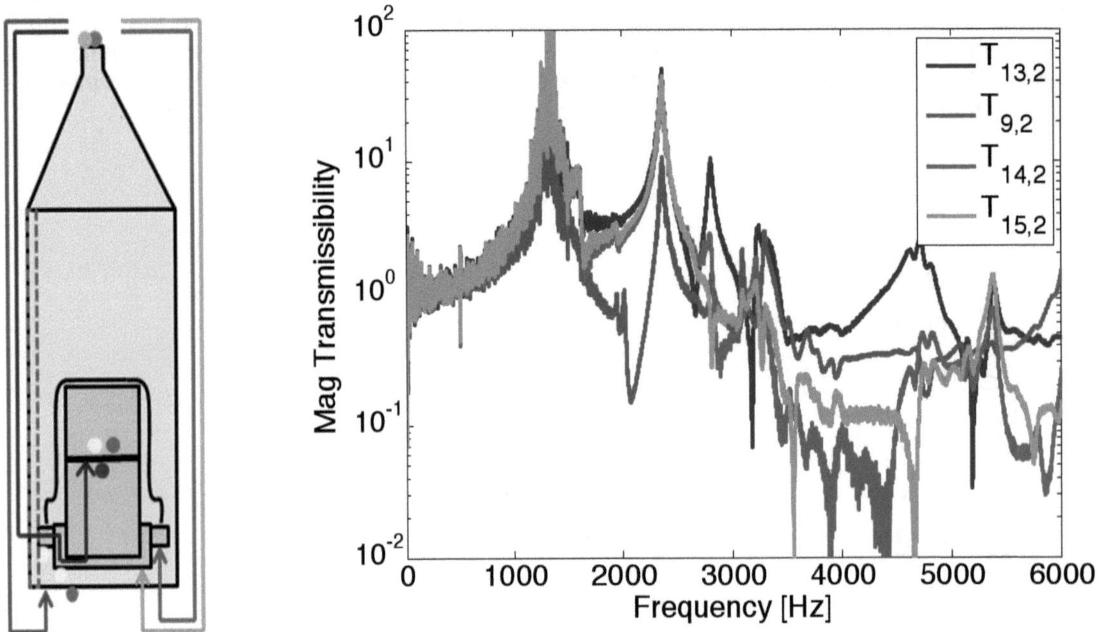


**Figure 9:** (a) Transmissibility pathways from tip to various locations, and (b) transmissibility response function magnitudes between axial motion on tip and motion on web (13), base of housing (9), retaining ring (14), and lock ring (15) showing differences in transmission along path to web.

To better interpret these transmissibility function plots, consider the two degree of freedom system shown in Figure 10. The nominally linear frequency response function relationship that describes the way in which low amplitude input forces in the frequency domain act on this system to produce output motions is given by,

$$\begin{Bmatrix} X_1(\omega) \\ X_2(\omega) \end{Bmatrix} = \begin{bmatrix} H_{11}(\omega) & H_{12}(\omega) \\ H_{21}(\omega) & H_{22}(\omega) \end{bmatrix} \begin{Bmatrix} F_1(\omega) \\ F_2(\omega) \end{Bmatrix}$$
$$= \frac{1}{(K_2 - M_2\omega^2)(K_1 + K_2 - M_1\omega^2) - K_2^2} \begin{bmatrix} K_2 - M_2\omega^2 & K_2 \\ K_2 & K_1 + K_2 - M_1\omega^2 \end{bmatrix} \begin{Bmatrix} F_1(\omega) \\ F_2(\omega) \end{Bmatrix} \quad (1,2)$$

where the mass and stiffness parameters are assigned as illustrated in Figure 10. For simplicity, damping has been removed from this model. The roots of the denominator polynomial given above are equal to the undamped natural frequencies of the model. If the ratio of the frequency response functions, $H_{12}$ and $H_{22}$, is computed when $F_2$ is applied for $F_1$=0, the transmissibility function, $T_{12(2)}$, is obtained:

$$T_{12(2)} = \left.\frac{X_1(\omega)}{X_2(\omega)}\right|_{with\,F_2} = \frac{H_{12}(\omega)}{H_{11}(\omega)} = \frac{K_2}{K_1 + K_2 - M_1\omega^2} \quad (3)$$

Note that $T_{12(2)}$ has a peak at the frequency $\omega = \sqrt{(K_1 + K_2)/M_1}$, which is the "internal" resonant frequency that occurs if $M_2$ is a fixed point. This frequency also corresponds to an anti-resonant frequency of the driving point frequency response function, $H_{22}$ (see Eq. (2)). It can be concluded that an anti-resonance in this two degree of freedom model is equivalent to an internal resonance of a substructure of the overall system. Likewise, the anti-resonance at 1400 Hz in the driving point frequency response function at the tip of the specimen (see Figure 6) produces a peak in the transmissibility function (Figure 9) – this peak corresponds to an internal resonance of the canister as if the tip was a fixed point. This interpretation can be used for all peaks in the transmissibility functions presented in this paper.

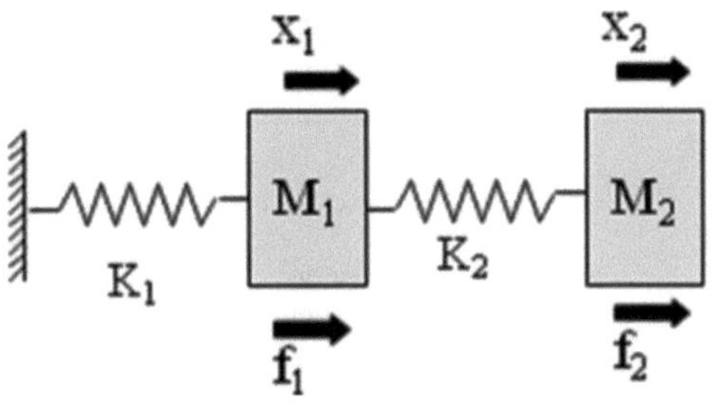


**Figure 10**: Two degree of freedom system.

Figure 11(a-d) shows the transmissibility magnitudes for the pathways illustrated in Figure 9(a) for hand tightened (• • • ) and 250 ft-lb tightened ( ▬ ) conditions. Note that all of the functions exhibit the same characteristics regardless of preload below roughly 1400 Hz. This result is consistent with the previously discussed findings, which concluded that the specimen exhibits primarily rigid body motions below 1400 Hz. Also note that near the resonant frequency at 2210 Hz, all of the transmissibility functions exhibit the same shift in frequency – this shift is due to the mass loading that the specimen housing introduces as it serves as a boundary condition to the oscillating canister. This peak represents the resonance that would occur if the housing was fixed and the canister as to oscillate relative to the housing on the interface between the retaining ring and the housing. As one progresses from Figure 11(a) to Figure 11(d) from top to bottom, it is possible to gain insight into the varying effects under preload of the interfaces and the web flexible behavior on the dynamic response. Above 2210 Hz from 2210 to 4000 Hz, the three interfacial motions appear to dominate the response leading to significant dependence on preload (which should also lead to large changes due to excitation amplitude when nonlinearity is considered). Also, the web flexible body behavior appears to dominate in the frequency range from 4000-6000 Hz for the two preload conditions – by tightening the lock ring, the boundary condition stiffness for the web resonant frequency was increased resulting in an effective increase in the resonant frequency of the web.

Figure 12 is a plot of the transmissibility functions between axial motions at the applied modal impact location and web axial motions for various impact locations on the tip and then on the nose in four different quadrants. First, notice that by moving the impact location to the nose, the peak in the transmissibility function now occurs near 2000 Hz as opposed to the lower peak at 1400 Hz that occurs for an impact at the tip location. This peak moves because of the significant change in density of the material beneath the impact site. Second, notice that from 3500 to 6000 Hz, impact forces applied to the nose on its side transmit better to the web than the impact force applied at the tip of the specimen. This result suggests that the transmission of force to the web is amplified for impacts applied on the nose as opposed to the tip.

In addition to understanding sources of variability due to changes in preload across the lock ring, sources of measurement variability were also ascertained. Figure 13 shows a comparison of the frequency response functions for the top modal triaxial accelerometers and base shock triaxial accelerometers mounted to the web of the canister. The shock accelerometer would be more appropriate in practice. Note that the two sets of z and y frequency response functions match reasonable well at the peaks near 2200, 2750, 3200, and 3500 Hz – the noise on the shock accelerometer channels is evident due to the low response amplitudes that are being recorded from the modal impacts. Between the peaks, there would be greater variability in the dynamic measurement leading to errors in the estimated force state of the specimen if these sensors were used for that purpose. Likewise, above 5000 Hz, the z direction accelerations match but the y direction acceleration readings do not. This mismatch may be due in part to the complexity of the modeshapes of the web in that frequency range.

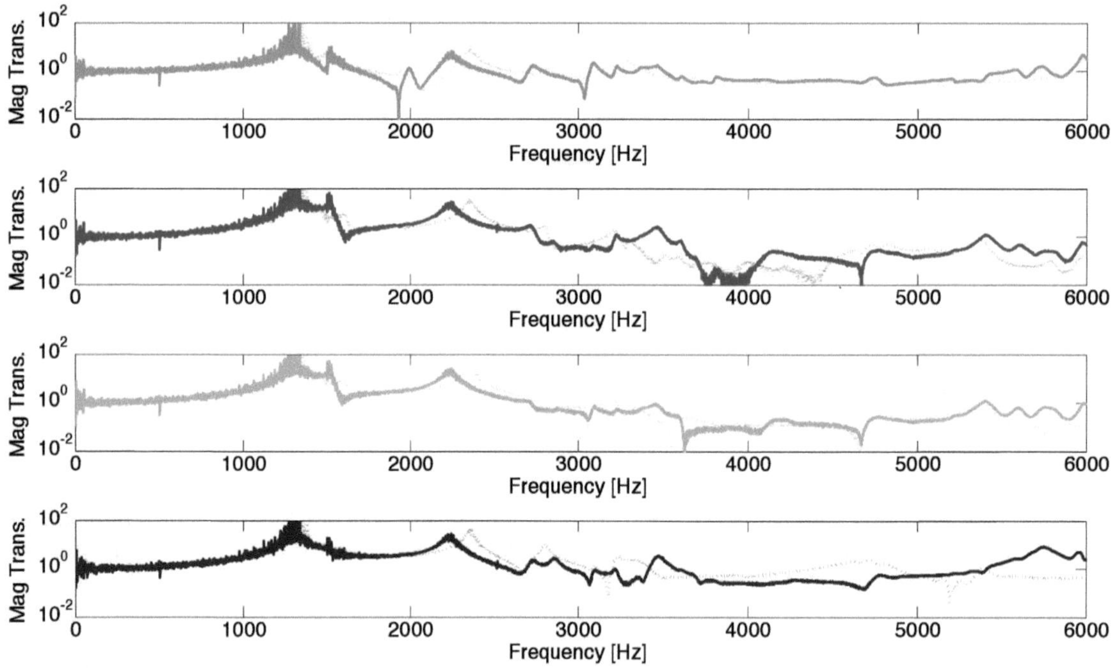


**Figure 11:** Transmissibility response magnitudes between (a) tip and web, (b) tip and housing base, (c) tip and retaining ring, and (d) tip and lock ring showing contributions of various interfaces for two preloads (hand tight, • • • ; and 250 ft-lb tight, ▬ ).

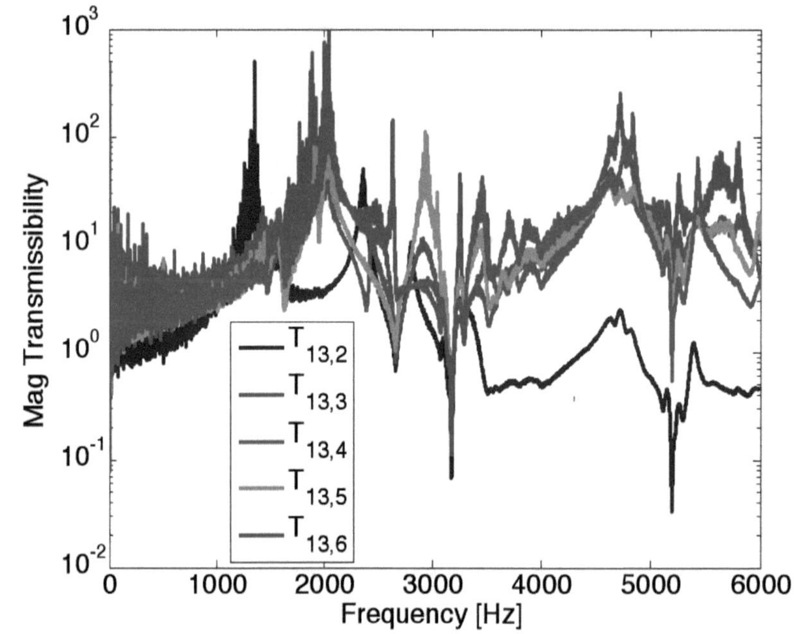


**Figure 12:** Transmissibility response functions from tip axial to web axial motions ( ▬ ) and from various locations on the nose in radial direction to web axial motions ( ▬ , ▬ , ▬ , ▬ ).

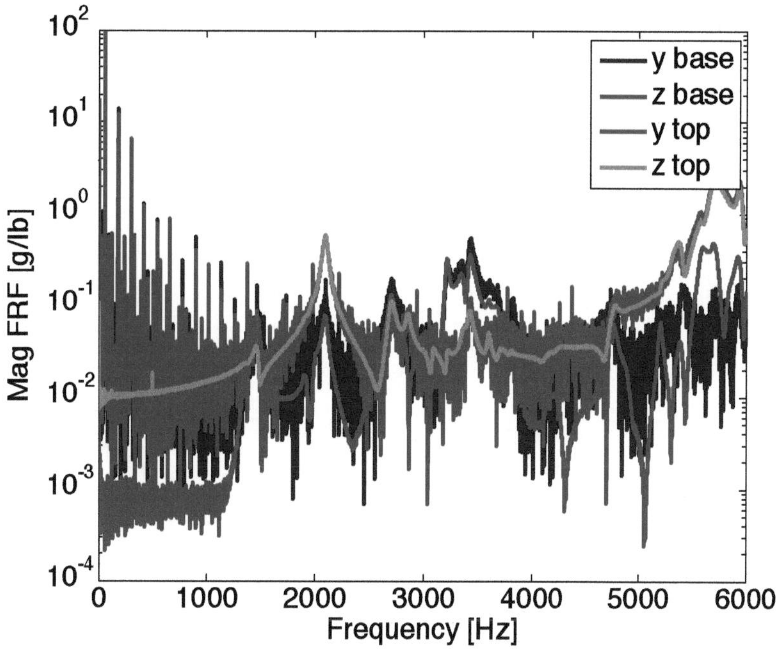


**Figure 13:** Frequency response functions at base and top of web inside canister showing match between modal (y ▬ ,z ▬ ) and shock (y ▬ ,z ▬ ) triaxial accelerometer channels near resonant frequencies.

The excitation amplitude was then varied using a larger, lower sensitivity modal impact hammer applied to the tip of the specimen with the 250 ft-lb preload applied to the lock ring. Figure 14 shows the frequency response functions between this applied force and the web acceleration response in the axial direction as a function of amplitude. Note that all of the frequency response functions approach 0.01 g/lb as the frequency approaches zero as in the previous measurements because this low frequency amplitude is equivalent to 1/m, where m is the specimen mass (approximately 100 lb). Also note that for increasing impact amplitude, the two strong resonances at 1400 and 2200 Hz decrease as do the three weaker resonances above 4000 Hz. These decreases can be explained by considering the linear modal deflection shapes that were explained previously. Recall that the modes of vibration at 1400 and 2200 Hz correspond to synchronous motions of the canister, lock ring, and retaining ring relative to the housing across the interface between the housing and the retaining ring. In other words, the two lower frequency modes experience a softening effect for large impact amplitudes as the canister oscillates relative to the housing across that threaded interface. Likewise, the higher frequency response exhibit softening behavior for the same reason, because the canister oscillates relative to the housing. In the mid-frequency range from 2200 to 4000 Hz, there are very complex nonlinear behaviors due to the relative, asynchronous motions between the canister, lock ring, and retaining ring as explained previously. Due to these complexities, it is presently unclear if and why the third strong peak shifts outward in frequency as the amplitude of excitation is increased.

## Section IV: Conclusions

The nominally linear modal behavior of a specimen consisting of a steel cylindrical housing with an internal canister that were connected through a threaded interface was experimentally analyzed. It was shown using the quadrature component of the frequency response functions that below 1400 Hz, the specimen behaved like a rigid body. Between 1700 to 2200 Hz, the canister oscillated on the threaded interface between a retaining ring and the housing. In this frequency range, the components separated by threaded interfaces were synchronous in their motions. Between 2200 and 4000 Hz, there were strong asynchronous motions that took place across these interfaces. By analyzing the transmissibility functions, it was shown that peaks in the transmissibility function corresponded to internal resonances with artificially imposed boundary conditions. It was also shown that the transmissibility functions could be used to identify which interfaces were resulting in the most change in the forced response due to changes in preload of the lock ring. The effects of increasing amplitude of excitation were also considered by applying modal impacts at various mean force levels. It was shown that softening nonlinear behaviors were observed below 2200 Hz and above 4000 Hz due to the threaded interfaces when synchronous motions governed the responses of the internal components within the specimen; however, complex nonlinear behaviors occurred due to asynchronous motions among these internal components in the mid-frequency range from 2200 to 4000 Hz.

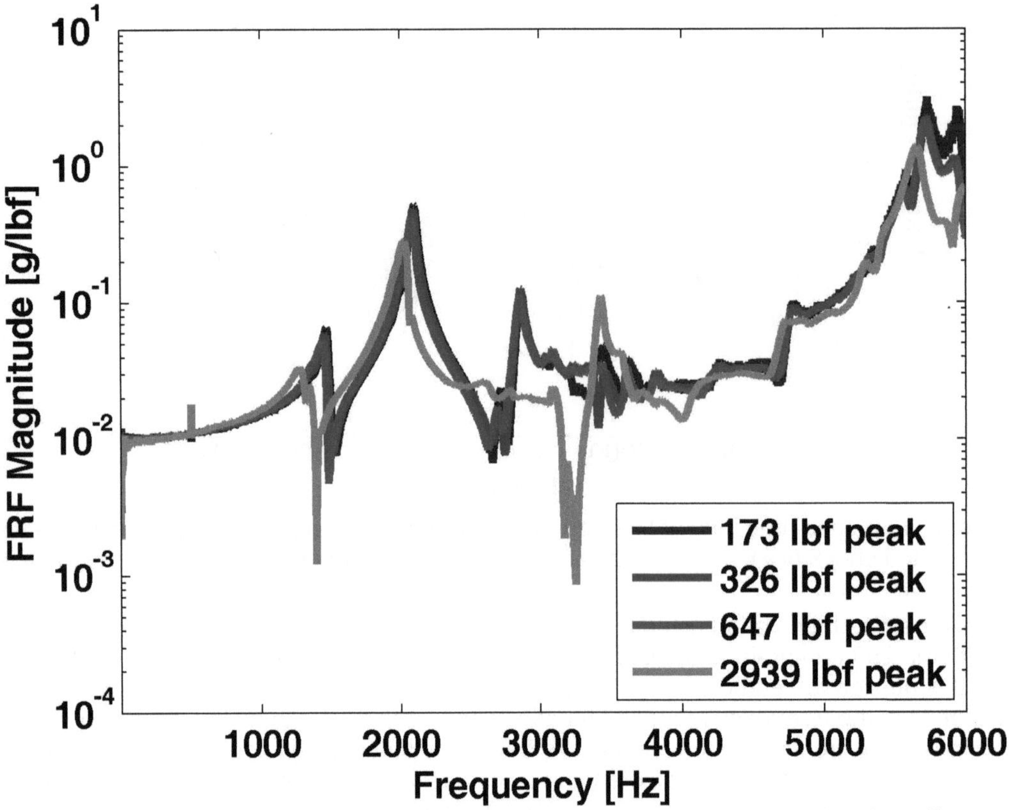


**Figure 14:** Frequency response functions between tip force and web response showing effects of increase in mean impact force.

## Section V: References


Ibrahim, R. A., and Pettit, C. L., "Uncertainties and Dynamic Problems of Bolted Joints and Other Fasteners," 2005, Journal of Sound and Vibration, Volume 279, Issues 3-5, 21, pp. 857-936.

Doebling, S., Schultze, J., and Hemez, F., "Validation of the Transient Structural Response of a Threaded Assembly," 2002, AIAA-2002-1644, 43rd AIAA/ASME/ASCE/ASC Structures, Structural Dynamics, and Materials Conference, Denver, CO.

Hess, D. P., and Sudhirkashyap, S. V., "Dynamic Analysis of Threaded Fasteners Subjected to Axial Vibration," 1996, Journal of Sound and Vibration, Volume 193, Issue 5, pp. 1079-1090.

Rashquinha, I. A., and Hess, D. P., "Modelling Nonlinear Dynamics of Bolted Assemblies," 1997, Applied Mathematical Modelling, Volume 21, Issue 12, pp. 801-809.

Proceedings of the IMAC-XXVIII
February 1–4, 2010, Jacksonville, Florida USA


# Experimental Study on Parametric Anti-resonances of an Axially Forced Beam

Horst Ecker and Ingrid Rottensteiner

Vienna University of Technology, Institute of Mechanics and Mechatronics,
Wiedner Hauptstrasse 8-10/E325/A3, 1040 Vienna, Austria

## ABSTRACT

Long and flexible beam structures are in use in various fields of engineering. Vibration problems are common with such structures due to high flexibility and low damping. Therefore, new ideas to enhance the damping of lateral vibrations are of interest. In this experimental study, a slender non-metallic cantilever beam is subject to periodic forcing along the beam axis. The force is generated by a wire, which is attached to the free end and runs to the clamped end of the beam, where it is connected to a piezoelectric stack actuator. Operating in an uncontrolled mode, the actuator extends proportionally to an input signal that results in a force according to the total axial stiffness of the structure. By applying a pre-tension to the beam, the axial force at the tip of the beam can be modulated in time by the actuator and the system becomes parametrically excited. For modulation frequencies near non-resonant parametric resonances it is known from theoretical studies that such a system will exhibit increased damping properties. This study presents measurement results on such a system, verifying theoretical predictions and confirming first measurements. It is shown that the attenuation of free vibrations can be increased significantly by the proposed method. While the appropriate frequency must be met within rather narrow limits to achieve best results, the method is relatively insensitive to a distortion of the harmonic input signal and also quite tolerant regarding amplitude fluctuations.

## 1. Introduction

During the last years there has been an increasing interest in mechanical structures with parametric excitation (PE). This interest is fuelled by the unusual behaviour that such systems exhibit. Parametrically excited systems may become unstable within several small ranges of the excitation frequency, leading to large vibration amplitudes. This characteristic may be dangerous for a structure and, in general, operation within the region of instability must be avoided. However, one may also take advantage of this behaviour, as this was proposed by Shaw et al. in [1], where parametric resonances, as opposed to the usual linear resonance, are used in MEMS oscillators for frequency selection.

Another feature of parametrically excited system is an increased level of damping at certain PE-frequencies. This phenomenon was first reported and analysed by Tondl in [2]. In his paper it is shown that intervals of enhanced stability may exist in a PE-system. In accordance with the commonly used terminology, which uses the expression "parametric resonance" for intervals of instability, such stability intervals are therefore called *non-resonant parametric resonances*. In the past, mainly lumped mass systems have been investigated. However, in [3] and in [4] the axially loaded cantilever beam was investigated numerically and enhanced vibration suppression was found for the axial force frequency near the anti-resonance frequencies. An additional study and a numerical comparison with a closed-loop control scheme was reported in [5]. Most recently, these theoretical results were confirmed in an experimental study [6].

The present paper is an extension of the work presented in [6]. One of the goals was to simplify the test stand, another one to investigate a non-metallic low-damping material for the cantilever beam. In the following, the basic principle of the test stand and the design are explained, followed by a short section on parametric resonances. The main part of the paper is devoted to the discussion of the measurement results.

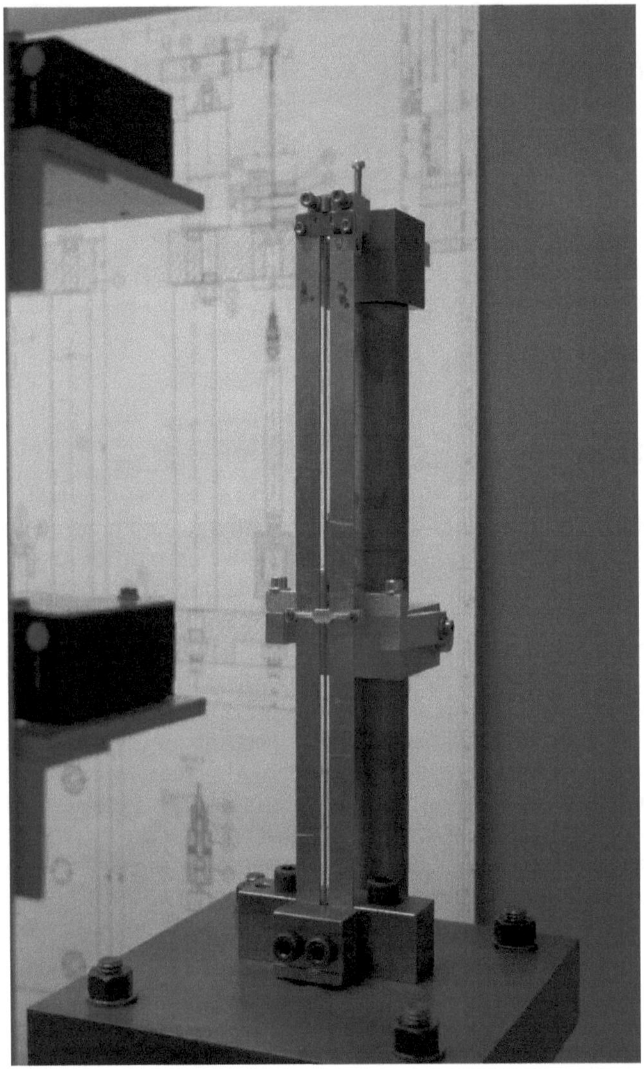

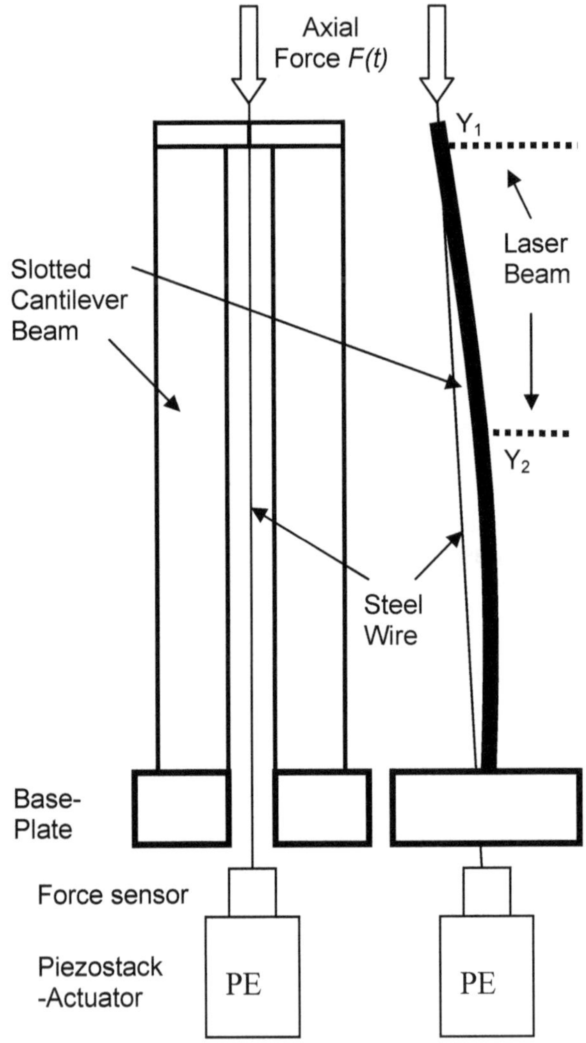


**Figure 1**: (Left) Picture of the test stand as used in [6] with Aluminum cantilever beam. (Right) Schematic of the test stand, showing the principle of generating an almost axial force by a tension wire and a piezoelectric actuator to create parametric excitation (PE). Dotted lines symbolize the measurement points for the beam deflection by a laser measurement system.

## 2. Mechanical principle of the test stand

Figure 1 shows a photograph of the test stand and a schematic diagram of the mechanical system. It consists of a slender and uniform cantilever beam with length $L$, Young's modulus $E$, cross section $A$, mass density $\rho$ and bending stiffness $EI$. In order to run a wire from the tip of the beam to the clamped end, the beam has an axial slot which ends near the tip, see Figure 2. A force

$$F(t) = F_0 + F_p(t), \quad F_p(t) = F_A \sin(t) \equiv F_A \sin(t + T), \tag{1}$$

acts at the tip of the beam. $F_0$ represents a constant preload force and $F_p(t)$ is the time-periodic component of the tip force.

Using a reduced order model the dynamical behaviour of the system can be described by its first and second bending mode. Accordingly, the set of differential equations is

$$\mathbf{M}\,\mathbf{q}'' + \mathbf{C}\,\mathbf{q}' + \mathbf{K}\,\mathbf{q} + F(t)\,\mathbf{U}\,\mathbf{q} = \mathbf{0}, \tag{2}$$

for a generalized displacement vector

$$\mathbf{q}(t) = \{ X_1(t) \, , X_2(t) \}^T \tag{3}$$

and (2x2) matrices **M** (mass matrix), **C** (damping matrix), **K** (stiffness matrix) and **U** (geometry matrix). Note that the system is homogeneous. The expression $F(t)$ **U** represents the time-periodic contribution of the tip force in the system. Since the physical unit of this expression is that of a stiffness parameter, the system exhibits a parametric stiffness excitation. A comprehensive discussion of the modelling of this system, including a FE-approach, is found in [3], [4] and [5].

## 3. Parametric instabilities – "Resonance" and "Anti-resonance" frequencies

Parametric instabilities occur at multiples, fractions and combinations of the natural frequencies of the system, see e.g. [8]. From Eq.(2) the time-constant system without damping can be derived

$$\mathbf{M}\,\mathbf{q}'' + \mathbf{K}\,\mathbf{q} + F_0\,\mathbf{U}\,\mathbf{q} = \mathbf{0}. \tag{4}$$

Note that the pre-load (tension) $F_0$ of the system needs to be retained in the equation, since it affects the natural frequencies. By a standard procedure it is possible to calculate the eigenvalues $\lambda_i$ and the natural frequencies

$$\omega_1 < \omega_2 \tag{5}$$

of system Eq.(4). With Eq.(5) the parametric primary resonances $\eta^{pr}_{1/n}$ , $\eta^{pr}_{2/n}$ and the parametric combination resonance $\eta^{c}_{s/n}$ of order $n$ are defined as follows:

$$\eta^{pr}_{1/n} = (2\,\omega_1)\,/\,n\,; \quad \eta^{pr}_{2/n} = (2\,\omega_2)\,/\,n\,; \quad \eta^{c}_{s/n} = (\omega_1 + \omega_2)\,/\,n\,; \qquad (n = 1,2, \ldots) \tag{6}$$

Near these parametric resonance frequencies intervals of instability have to be expected. In general, significant instabilities only occur for order n=1 and occasionally for order n=2. Higher orders (n > 2) are very rarely observed. In a theoretical study by Chen and Yeh [7] instability maps for a beam system have been calculated. Also [4] shows the respective instability areas, based on a detailed FE-model.

The parametric combination resonance frequency of the difference type, however,

$$\eta^{c}_{d/n} = (\omega_2 - \omega_1)\,/\,n \qquad (n = 1,2, \ldots) \tag{7}$$

is a non-resonant parametric resonance, provided that the axial force at the tip of the beam has a fixed direction. Again, only orders $n = 1$ and $n = 2$ are significant. In the vicinity of the anti-resonance frequency $\eta^{c}_{d/1}$ not only no instability is observed (see Chen and Yeh [7]), in fact enhanced damping can be noted if certain criteria are fulfilled [3], [5]. It is the purpose of the test stand to prove that vibration damping can be achieved by axial forcing of the beam.

## 4. Design of the test stand

Figure 1 and 2 show photographs of the test stand. The non-metallic cantilever beam is made of polycarbonat (Macralon®) and is mounted to the base plate of the supporting structure in an upright position. See the Appendix for characteristic data of the test stand. The beam can perform unrestricted bending vibrations. A steel wire is attached to the upper end of the beam and runs in a slot from the tip to the root of the beam. The string passes through the base plate and is connected to the piezoelectric stack actuator. This device is used to create an almost vertical force at the tip of the beam. Compared to Ref. [6] the mounting of the actuator has been modified such that the actuator now cannot follow a lateral motion of the wire. Instead, it is rigidly attached to the foundation of the test stand.

A voltage applied to the piezoelectric actuator results in an extension and is provided by a piezo-amplifier. The control signal to the amplifier is generated by a host computer via the analog output of a controller card. A force sensor is installed between the actuator and the bottom end of the string. The signal from this sensor is only used for adjusting the pre-load $F_0$ and to record the applied force to the beam. In [6] full force control was implemented based on the measured force signal. Now, in order to investigate a simplified system, the actual applied force was not controlled. The test stand was excited only by a feed forward voltage to the piezoelectric actuator, which resulted in a certain force due to the elongation of the actuator and the overall stiffness of the system. Therefore, also the actual excitation force signal could and did deviate from the ideal harmonic function. It was one of the goals to test how sensitive the system is to such imperfections.

Beam vibrations were picked up at the tip of the beam and at midspan position by a laser distance measurement system. Both signals were only recorded along with the force signal, but not used for any kind of closed-loop control. Excitation amplitude and frequency were set before each experiment and left unchanged during the run. All tests were conducted according to the following procedure. The pre-load was adjusted by the tension system at the upper end of the steel wire. Then the beam was deflected at the tip by the deflection device. The actuator was moved forward (elongated) half of its maximum travel. Then the beam was released and parametric excitation by harmonic vibration of the actuator was started at the same time.

**Figure 2**: (Left) Piezoelectric actuator situated below the base plate with force sensor and steel wire attached on top. (Right) Tip of the slotted Makralon® cantilever beam with screw to adjust the pre-load of the steel wire. The deflection device is in the open position.

## 5. Measurement results

After setting the preload force or making adjustments to it, the natural frequencies of the beam system were measured by recording the deflection signals after an impact excitation and by analysing the signals with an FFT-analyser. A typical result can be seen in Figure 3, showing two major and distinct peaks. The first bending frequency can be identified at about 7,5 Hz and the second one at about 52 Hz. The calculation of respective PE-frequencies was always based on the measured natural frequencies. It turned out that this was a very reliable method and that the predicted parametric frequencies were in close agreement to the measurements.

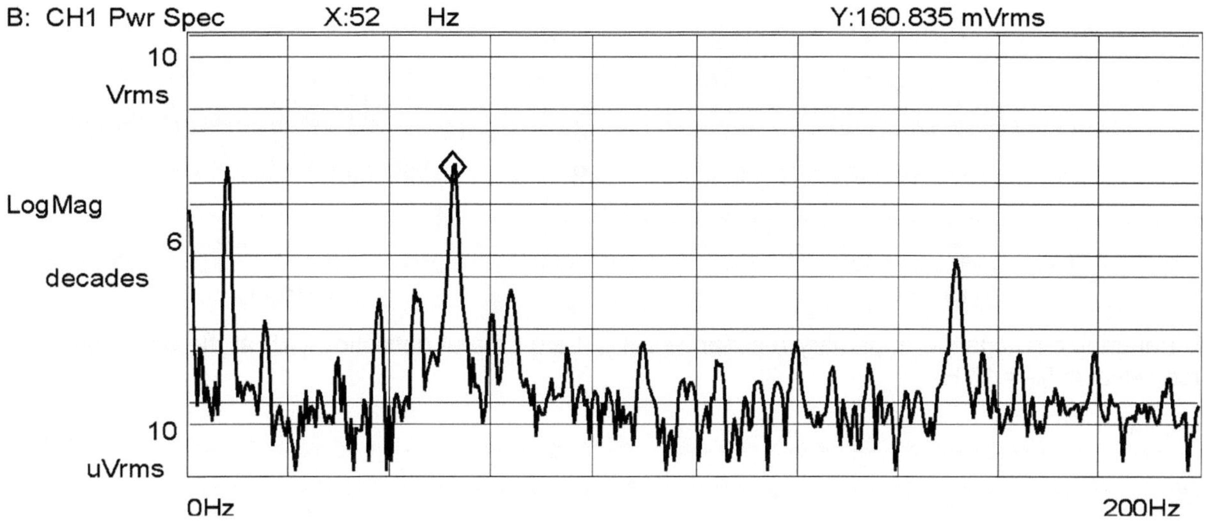


**Figure 3**: Typical FFT-spectrum of the lateral vibration signal of the beam after an impact excitation, showing the first natural frequency of the beam at about 7,5 Hz and the second natural frequency at about 52 Hz.

First, series of measurements are shown to demonstrate the effect of parametric resonances on this test stand. In the following Figures 4 to 7, the diagram on the left hand side is the vibration signal at measurement position $y_1$ at the tip of the beam. The diagram to the right is the actual force signal as measured by the force sensor. Note that the preload force is not shown and therefore the signal starts at zero. Also the force signal is uncontrolled. The voltage provided to the piezo-amplifier is a harmonic signal of constant frequency and amplitude. Due to the reaction of the vibrating beam, the actual force signal is distorted. The force amplitude varies and also other frequencies do appear in the signal. The measurements were carried out such that the beam was released from its deflected position of 1.5 mm at the tip of the beam and at the same time the parametric excitation was turned on. Note that the initial deflection of the beam is close to the first bending mode. Therefore, first mode vibrations are the main vibration signal at the beginning of each run.

Figure 4 shows the result when the parametric excitation frequency $f_{PE}$ = 15 Hz is set to the primary resonance frequency $\eta^{pr}{}_{1/1} = 2\omega_1$. After about one second of free vibrations the amplitudes start to increase and a strong parametric resonance begins to develop. After two seconds the amplitudes have grown to almost 3 mm and the geometric limitations of the test rig prevent even larger amplitudes. The frequency of the force signal is the frequency of the parametric excitation. Starting from a clean harmonic signal, both instability and non-linearity influence the force signal.

In Fig. 5 the parametric excitation frequency $f_{PE}$ = 104 Hz is set to the primary resonance frequency $\eta^{pr}{}_{2/1} = 2\omega_2$. It takes the system about 2 seconds to develop significant second mode vibrations. The force signal also clearly shows the beginning of the instability.

With the test stand it was also possible to demonstrate the combination frequency instability $\eta^{c}{}_{s/1} = (\omega_1 + \omega_2)$ at $f_{PE}$ = 60 Hz. Figure 6 shows the respective results. At this frequency, also the steel wire performed large vibrations and caused the intermittent limitation of the beam vibrations. The force signal also reflects the violent vibrations of the system.

Finally, the system was excited at the anti-resonance at $\eta^{c}{}_{d/1} = (\omega_2 - \omega_1)$ at $f_{PE}$ = 44.75 Hz. The results in Fig. 7 show that the first mode vibration signal decreases rapidly, while second mode vibrations grow temporarily after about 2 seconds, and then vanish again. The force signal is an almost steady signal with only some modulations. Clearly, there is no instability near this frequency. In Fig. 10 the deflection signal at both measurement points is shown over a period of 8 seconds and the enhancement of damping is demonstrated.

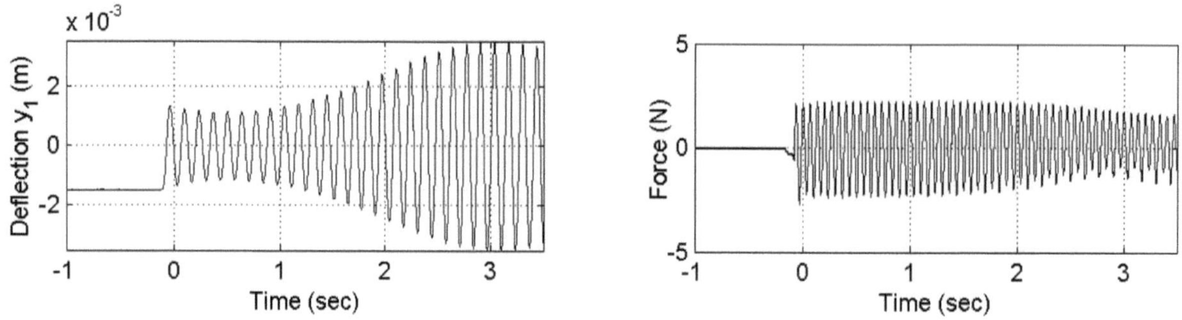


**Figure 4**: Parametric resonance at primary resonance $2\omega_1$. (Left) Lateral deflection $y_1(t)$ at the tip of the beam, (right) force signal at $f_{PE}$ = 15 Hz

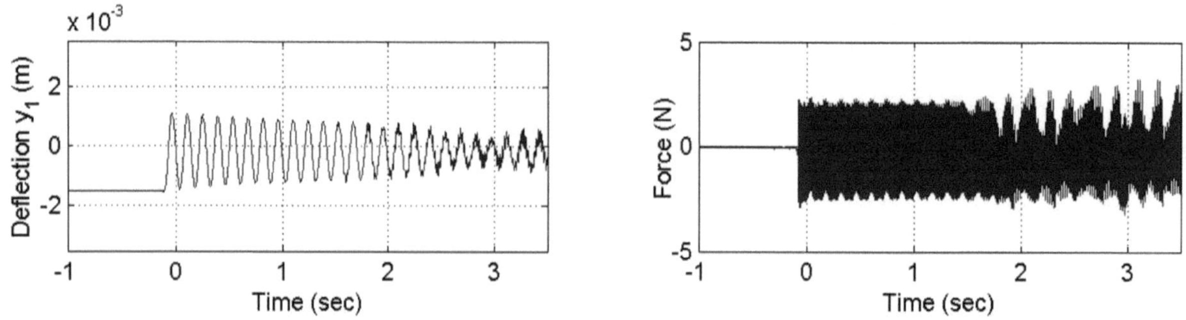


**Figure 5**: Parametric resonance at primary resonance $2\omega_2$. (Left) Lateral deflection $y_1(t)$ at the tip of the beam, (right) force signal at $f_{PE}$ = 104 Hz

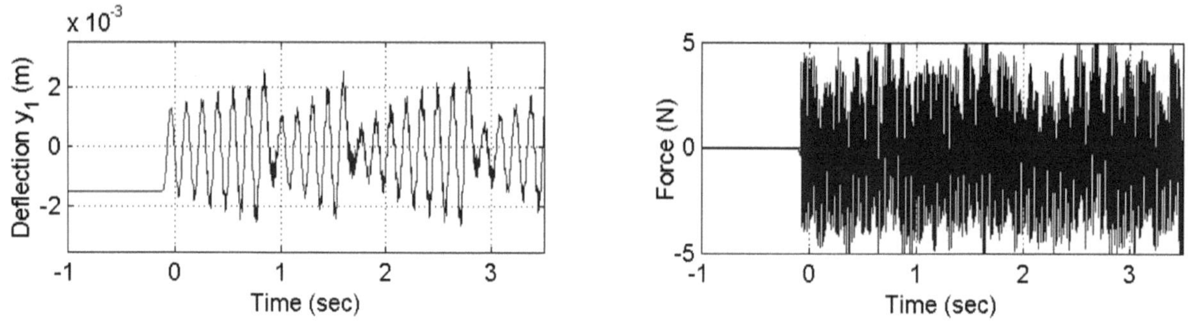


**Figure 6**: Parametric resonance at combination resonance ($\omega_1 + \omega_2$). (Left) Lateral deflection $y_1(t)$ at the tip of the beam, (right) force signal at $f_{PE}$ = 60 Hz

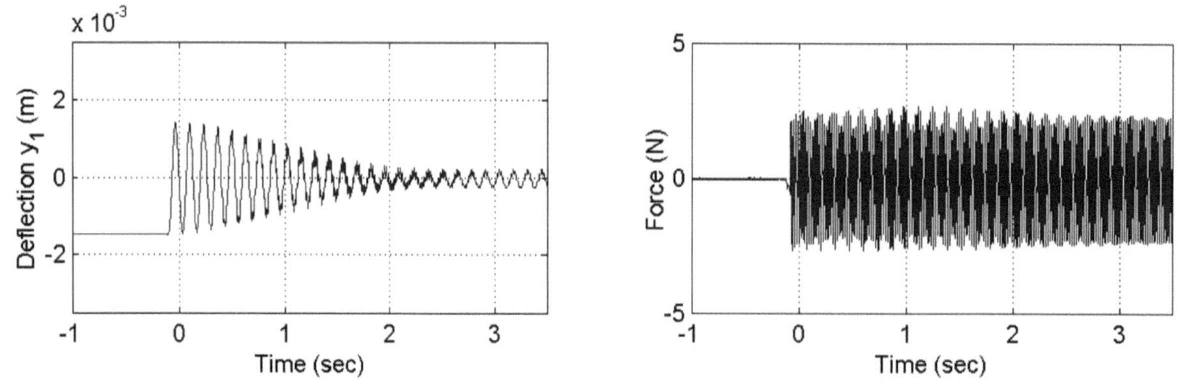


**Figure 7**: Parametric anti-resonance at combination resonance ($\omega_2 - \omega_1$). (Left) Lateral deflection $y_1(t)$ at the tip of the beam, (right) force signal at $f_{PE}$ = 44,75 Hz

To verify the increased damping capability of the system at the anti-resonance PE-frequency, a series of tests was carried out. The only parameter to be changed was the PE-frequency. A rather high resolution was selected and frequencies from 40 Hz to 50 Hz were tested. As a reference to be compared with, Fig. 8 shows the result when no parametric excitation was applied. First mode vibrations, initiated by the initial conditions, decay exponentially at a rather slow rate.

In Fig. 9 the result for $f_{PE}$ = 44,0 Hz is shown. Although the frequency is quite close to the optimal PE-frequency, the deflection signal $y_1$ looks very similar to the reference signal and no increased damping is apparent.

The result shown in Fig. 10 was obtained for $f_{PE}$ = 44,75 Hz and is the same as in the previous Fig. 7. In comparison with Figs. 8 and 9 one can see that, at this frequency, the vibrations of the beam are suppressed significantly faster with parametric excitation. It is remarkable that the rate of decrease is not constant. Between 0 and second 2 the vibrations at $y_1$ decrease very fast, but vibration signal $y_2$ even grows some, especially the high frequency content. Then the amplitudes of signal $y_1$ remain constant while for signal $y_2$ a decay of the high frequency amplitudes is noticed. After 6 seconds both signals have almost vanished, whereas the signal level in the case of no PE is still at about 25 percent of the initial level, see Fig. 8.

Finally, the last diagrams in Fig. 11 show the results for $f_{PE}$ = 46,0 Hz. Again, this result looks very similar to the reference signal, which shows that the interval of enhanced damping does not reach up to that frequency.

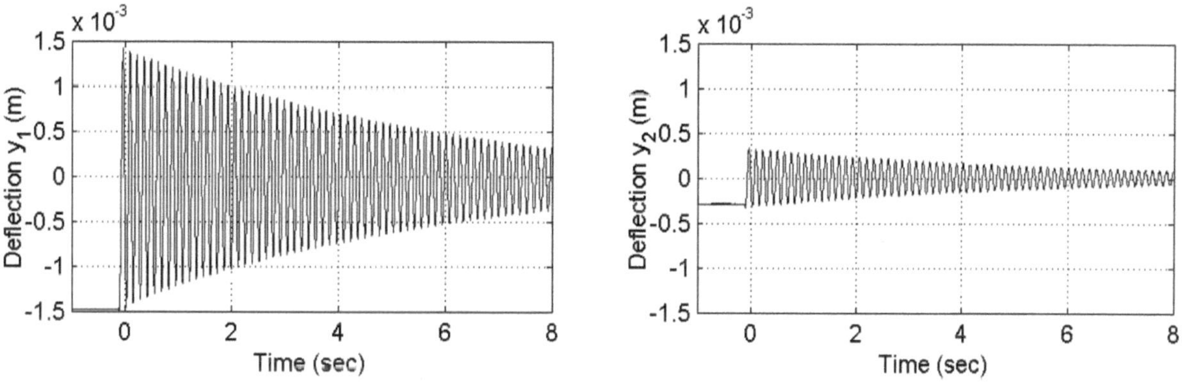


**Figure 8**: No PE-frequency $f_{PE}$ = 0 Hz. (Left) lateral deflection $y_1(t)$ at the tip of the beam, (right) lateral deflection $y_2(t)$ at the midspan position.

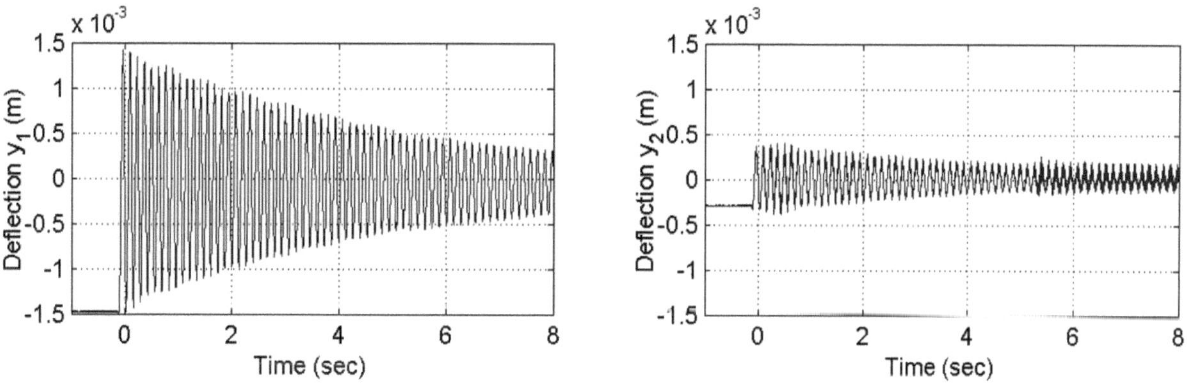


**Figure 9**: PE-frequency $f_{PE}$ = 44,0 Hz. (Left) lateral deflection $y_1(t)$ at the tip of the beam, (right) lateral deflection $y_2(t)$ at the midspan position.

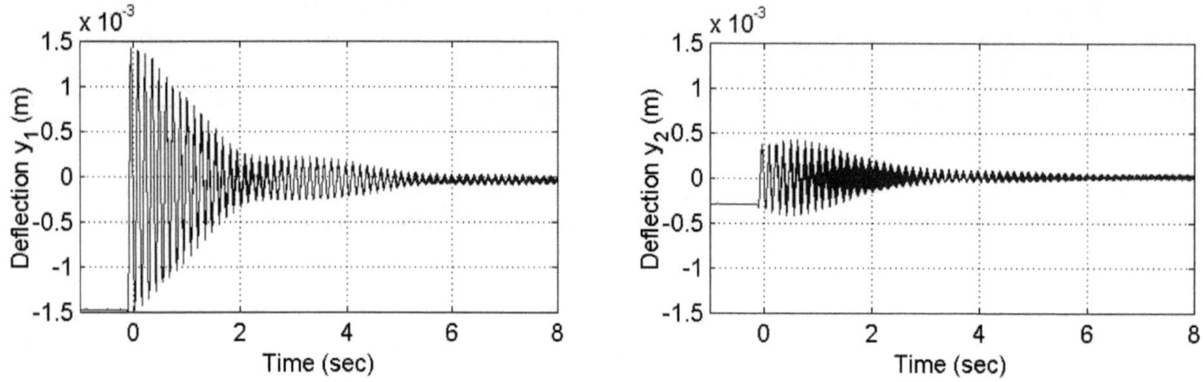


**Figure 10**: PE-frequency $f_{PE}$ = 44,75 Hz. (Left) lateral deflection $y_1(t)$ at the tip of the beam, (right) lateral deflection $y_2(t)$ at the midspan position.

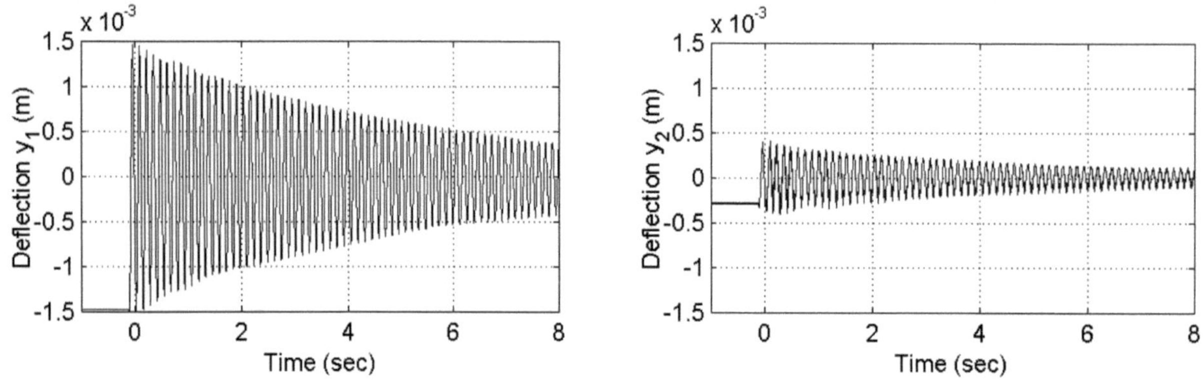


**Figure 11**: PE-frequency $f_{PE}$ = 46,0 Hz. (Left) lateral deflection $y_1(t)$ at the tip of the beam, (right) lateral deflection $y_2(t)$ at the midspan position.

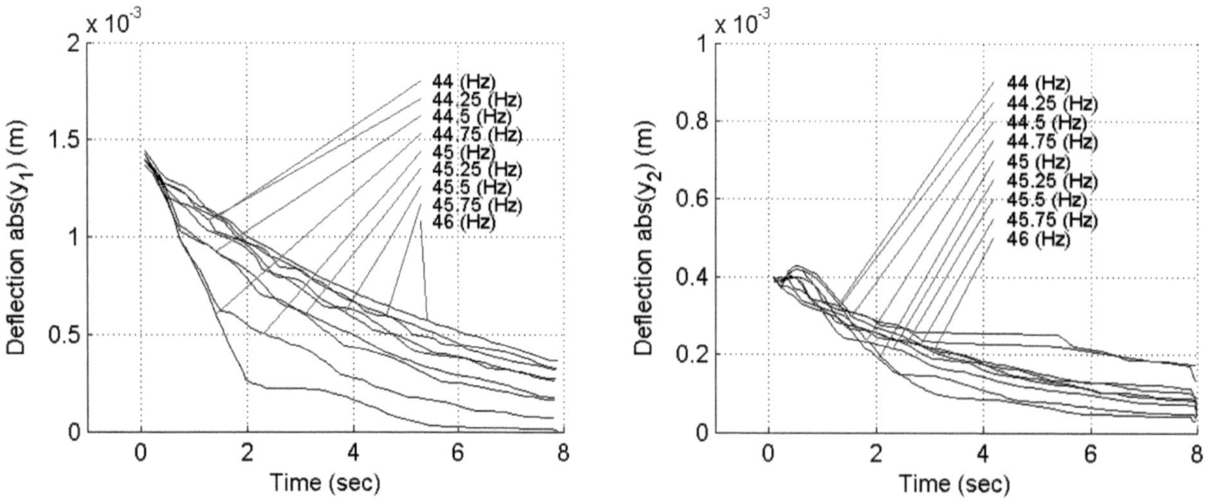


**Figure 12**: Envelope curves of vibration signals at various PE-frequencies. (Left) lateral deflection $y_1(t)$ at the tip of the beam, (right) lateral deflection $y_2(t)$ at the midspan position.

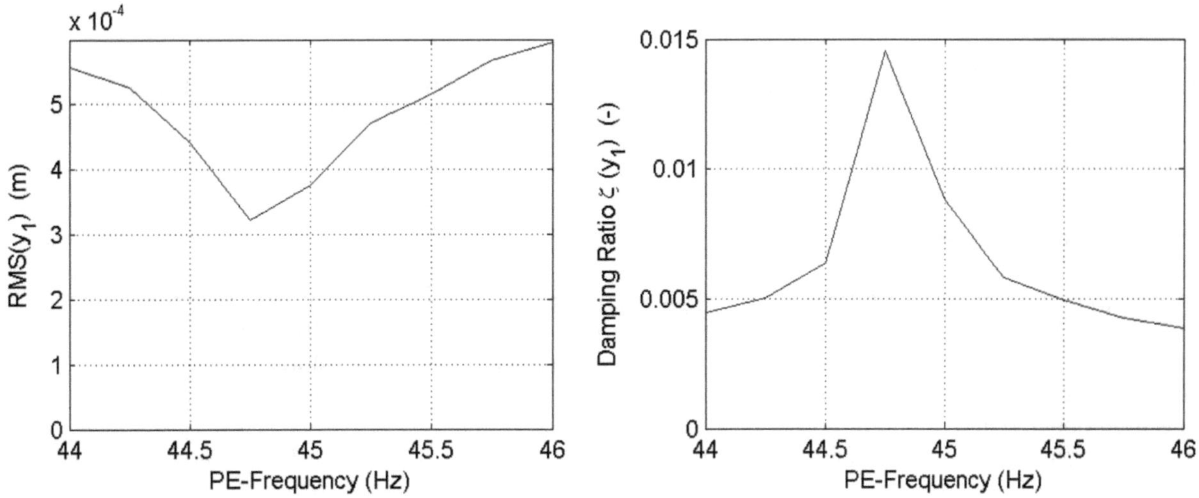


**Figure 13**: Characteristic parameters for vibration signals at various PE-frequencies. (Left) RMS value for signal $y_1(t)$ as shown in Figs. 9 -12, (right) equivalent damping ratio for signal $y_1(t)$ .

To facilitate the comparison of the many measurements, the envelope curves of the vibration signals obtained for PE-frequencies near the optimal PE-frequency were calculated and plotted. The diagrams in Fig. 12 show the respective results. The diagram for signal $y_1$ clearly shows the significantly increased damping at the PE-frequency $f_{PE}$ = 44,75 Hz. It is interesting to note that the signal $y_2$ at first exhibits a small increase of amplitudes, when a PE-frequency at or close to the optimal value is chosen. This behaviour is in agreement with earlier numerical [5] and experimental results [6].

Figure 13 shows two characteristic parameters of the deflection signal $y_1$. The left diagram presents a plot of the RMS value calculated for the signal within the time window as used in the previous diagrams. A smaller RMS value means that the signal did decay faster. It is worth to note that the smallest RMS value is almost half of the largest value within the frequency range of 2 Hz. On one hand this demonstrates the effectiveness of the method, on the other it also shows that the useful frequency range to achieve good vibration suppression is quite small.

Another parameter that is frequently used to characterize damping is the damping ratio. Based on the envelope curves, a curve fitting procedure was employed and averaged decay functions were calculated. Thereby, an equivalent damping parameter was found and the dimensionless damping ratio was calculated. The result of this procedure is shown in Fig. 13 on the right hand side. The equivalent damping ratio has a maximum value of almost 0.015 at the optimal PE-frequency $f_{PE}$ = 44,75 Hz. The gain in damping is about 3 times of that of the system without parametric excitation. This result is in very good agreement with earlier numerical and experimental results, see [5] and [6].

## 6. Conclusions

Measurements with a test stand to investigate enhanced damping achieved by parametrically exciting a beam were conducted. Parametric excitation is realized by an almost axial forcing of the cantilever beam. As far as the authors know, this is one of the very first experimental reports on parametric damping of a beam structure by axial forcing. The measured results clearly show the damping effect as predicted by analytical and numerical studies and are in very good qualitative agreement with previously obtained numerical results. On average, the structural damping of the cantilever beam can be enhanced by a factor of three, if parametric excitation is applied to the beam. Since parametric excitation, as used in this application, is based on open-loop control, it is not needed to measure the beam deflection. In this study a simplified and rigid mounting of the PE-actuator was used. Also, there was no force amplitude control implemented and only feed forward control of the actuator elongation amplitude was in use. The simplifications as used in this study, compared to a previous study, did not prevent the increase of damping as predicted from ideal mechanical models.

## Acknowledgment

The authors gratefully acknowledge the support of Dr. Thomas Pumhössel with regard to the test stand.

## Appendix

Characteristic data of the test stand:

| | | | |
|---|---|---|---|
| Effective width of the beam | 25 mm | Beam material | Polycarbonat Macralon® |
| Thickness of the beam | 4 mm | Mass density | 1200 kg/m3 |
| Free length L | 0.3 m | Young's modulus E | 2400 MPa |
| Diameter of steel wire | 1 mm | | |
| Measurement position $y_1$ | 0.3 m | Preload force | 5 – 8 N |
| Measurement position $y_2$ | 0.128 m | Harm. Force ampl. | 0 – 5 N (as measured) |

| | |
|---|---|
| 1st natural bending frequency | ≈ 7,5 Hz (dep. on preload force) |
| 2nd natural bending frequency | ≈ 52 Hz (dep. on preload force) |

## References

[1] Shaw, S.W., Turner, K.L., Rhoads, J.F. and Baskaran, R, 2003. "Parametrically Excited MEMS-Filter". In Proc. of IUTAM Symposium on Chaotic Dynamics and Control of Systems and Processes in Mechanics, (eds. G. Rega, F. Vestroni), Rome, Italy, 8-13 June, 2003.

[2] Tondl, A., 1998. "To the problem of quenching self-excited vibrations". *Acta Technica CSAV,* **43**, pp. 109–116.

[3] Ecker, H., Dohnal, F., and Springer, H., 2005. "Enhanced damping of a beam structure by parametric excitation". In Proceedings ENOC-2005, D. H. van Campen, M. Lazurko, and W. van den Oever, eds. Paper-Nr. 22-271.

[4] Petermeier, B., 2006. "Parametrically excited vibrations of a cantilever beam". MS Thesis, Vienna University of Technology, Vienna, Austria. (in German).

[5] Pumhössel, T. and Ecker, H., 2007. "Active damping of vibrations of a cantilever beam by axial force control". In Proc. of 2007 ASME International Design Engineering Technical Conference, ASME. DETC2007-34638.

[6] Ecker, H. and Pumhössel, T, 2009. "Experimental Results on Parametric Excitation Damping of an Axially Loaded Cantilever Beam". In Proc. of 2009 ASME International Design Engineering Technical Conference, ASME. DETC2009-86555.

[7] Chen, C.-C., and Yeh, M.-K., 1995. "Parametric instability of a cantilevered column under periodic loads in the direction of the tangency coefficient". *Journal of Sound and Vibration,* **183**(2), pp. 253–267.

[8] Cartmell, M., 1990. *Introduction to Linear, Parametric and Nonlinear Vibrations.* Chapman and Hall, London, UK.

Proceedings of the IMAC-XXVIII
February 1–4, 2010, Jacksonville, Florida USA


# Phase Resonance Testing of Nonlinear Vibrating Structures

**M. Peeters, G. Kerschen, J.C. Golinval**

Structural Dynamics Research Group, Department of Aerospace and Mechanical Engineering,
University of Liège, Chemin des Chevreuils 1 (B52), 4000 Liège, Belgium
Email: m.peeters, g.kerschen, jc.golinval@ulg.ac.be

## ABSTRACT

Modal testing and analysis is well-established for linear systems. The objective of this paper is to progress toward a practical experimental modal analysis methodology of nonlinear mechanical structures. In this context, nonlinear normal modes (NNMs) offer a solid theoretical and mathematical tool for interpreting a wide class of nonlinear dynamical phenomena, yet they have a clear and simple conceptual relation to the classical linear normal modes (LNMs). A nonlinear extension of force appropriation techniques is developed in this study in order to isolate one single NNM during the experiments, similarly to what is carried out for ground vibration testing. With the help of time-frequency analysis, NNM modal curves and their frequencies are then extracted from the time series. The proposed methodology is demonstrated using a simple numerical benchmark, which consists of a two-degree-of-freedom system with a cubic spring.

## 1. INTRODUCTION

In the virtual prototyping era, dynamic testing remains an important step of the design of engineering structures, because the accuracy of finite element predictions can be assessed. In this context, experimental modal analysis (EMA) is indubitably the most popular approach and extracts the modal parameters (i.e., the mode shapes, natural frequencies and damping ratios). The techniques available today for EMA are really quite sophisticated and advanced: eigensystem realization algorithm, stochastic subspace identification method, polyreference least-squares complex frequency domain method, to name a few.

While the common practice is to assume linear behavior, nonlinearity is a frequent occurrence in engineering applications and can drastically alter their behavior. For instance, in an aircraft, besides nonlinear fluid-structure interaction, typical nonlinearities include backlash and friction in control surfaces and joints, hardening nonlinearities in engine-to-pylon connections, and saturation effects in hydraulic actuators. Satellites are other examples of aerospace applications where nonlinearity may significantly impact the dynamic behaviour [1,2].

As reported in [3], a large body of literature exists regarding dynamic testing and identification of nonlinear structures, but very little work addresses nonlinear phenomena during modal survey tests. Interesting contributions in this context are [4,5]. In this paper, an attempt is made to extend EMA to a practical nonlinear analog using the nonlinear normal mode (NNM) theory. NNMs offer a solid and rigorous mathematical tool for analyzing nonlinear oscillations, yet they have a clear conceptual relation to the classical linear normal modes [6]. Another appealing feature of NNMs is that they are capable of handling strong structural nonlinearity. Following the philosophy of force appropriation, the proposed method excites the NNMs of interest, one at a time. Thanks to the invariance principle (i.e., if the motion is initiated on one specific NNM, the remaining NNMs remain quiescent for all time), the NNM modal curves and their frequencies of oscillation can be extracted directly from experimental data.

When used in conjunction with the numerical computation of the NNMs introduced in [7], the approach described herein leads to an integrated methodology for modal analysis of nonlinear vibrating structures.

## 2. THEORETICAL MODAL ANALYSIS (TMA)

A detailed description of NNMs and of their fundamental properties (e.g., frequency-energy dependence, bifurcations and stability) is given in [6]. In the present study, an NNM motion is defined as a (non-necessarily synchronous) periodic motion of the undamped mechanical system.

The approach followed here for the TMA targets the numerical computation of undamped NNMs of nonlinear structures discretized by finite elements. The numerical method relies on two main techniques, namely a shooting procedure and a method for the continuation of periodic solutions. A detailed description of the numerical algorithm is given in [7]. The NNMs are then obtained accurately, even in strongly nonlinear regimes, and in a fairly automatic manner.

The shooting method consists in finding, in an iterative way, the initial conditions and the period inducing an isolated periodic motion (i.e., an NNM motion) of the conservative system. To this end, the method relies on direct numerical time integration and on the Newton-Raphson procedure. Starting from some assumed initial conditions, the motion of the system at the assumed period is determined by numerical time integration. The Newmark scheme is considered herein. According to the difference between the initial and the final states of the system, the initial conditions and the period are then corrected iteratively using the Newton-Raphson algorithm.

A continuation method is combined with the shooting method to determine successive NNM motions at different energy levels. The so-called pseudo-arclength continuation method is used herein. Starting from a single NNM motion at a specific energy level, another NNM motion at a different energy level is computed using a predictor step and a corrector step. Using the LNMs as initial guess at low energy, the NNM computation is carried out for increasing energy levels.

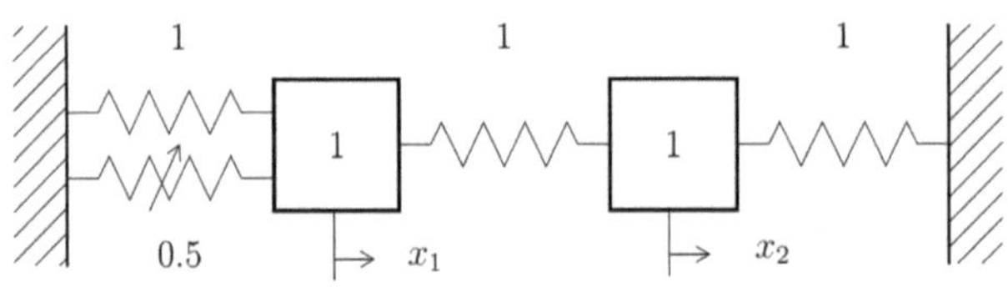


Figure 1: Two-degree-of-freedom system

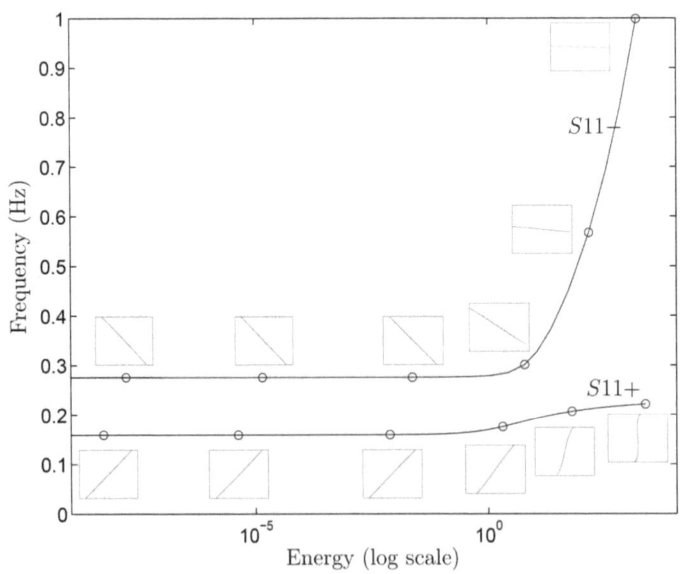


Figure 2: Frequency-energy plot of the 2DOF system. NNM motions depicted in the configuration space are inset. The horizontal and vertical axes in these plots are the displacements of the first and second DOFs, respectively.

One typical dynamical feature of nonlinear systems is the frequency-energy dependence of their oscillations. As a result, the modal curves and frequencies of NNMs depend on the total energy in the system. Due this dependence, the representation of NNMs in a frequency-energy plot (FEP) is particularly convenient. An NNM motion is represented by a point in the FEP, which is drawn at the fundamental frequency of the periodic motion and at the conserved total energy during the motion, which is the sum of the potential and kinetic energies. A branch, represented by a solid line, is a family of NNM motions possessing the same qualitative features.

For illustration, the 2DOF system with a cubic stiffness depicted in Figure 1 is considered. The underlying linear system possesses two (in-phase and out-of-phase) LNMs. The FEP, computed using the numerical algorithm, is shown in Figure 2. More complex systems can be considered without difficulty [8].

## 3. EXPERIMENTAL MODAL ANALYSIS (EMA)

There are two classical linear approaches for EMA. Phase separation methods excite several modes at once using either broadband excitation or swept-sine excitation covering the frequency range of interest. Phase resonance methods excite one mode at a time using multi-point sine excitation at the corresponding natural frequency. A careful selection of the shaker locations is required to induce single-mode behavior. This process is also known as normal-mode tuning or force appropriation.

Phase resonance methods have been used for decades, particularly in the aerospace industry (e.g., for ground vibration testing of aircrafts and modal survey of satellites). They provide a very accurate identification of the modal parameters, but they are time-consuming. In order to reduce the testing time, phase separation methods are now used in conjunction with phase resonance methods; these latter are reserved for modes that need a special attention (e.g., closely-spaced modes and important modes for flutter computations).

In this paper, an attempt is made to extend EMA to a practical nonlinear analog using the NNM theory. Realizing that the general motion of a nonlinear system cannot be expressed as a superposition of NNMs, it is not clear how individual NNMs can be extracted using phase separation methods. This is why our focus is on the extension of phase resonance testing to nonlinear structures.

The present section introduces a two-step approach that extracts the NNM modal curves and their frequencies of oscillation directly from experimental data. This method is summarized in Figure 3.

First, the method consists in exciting the system to induce single-NNM behavior at a specific energy level. To this end, an extension of force appropriation to nonlinear systems is developed. In particular, the phase lag of the response with respect to the applied force is used as an indicator to assess the quality of the appropriation. This first step, referred to as NNM force appropriation, is represented in the upper part of Figure 3.

The second step turns off the excitation to track the energy dependence of the NNM of interest using the free damped response resulting from the appropriated NNM motion. A schematic representation is presented in the lower part of Figure 3. During this NNM free decay identification, the frequency-energy dependence may be extracted using time-frequency analysis. More details on the two steps of the procedure are given in the next sections.

The methodology therefore consists in extracting one NNM at a time from the time series by means of this two-step procedure. In order to relate the forced and free responses of the damped system to the NNMs of the underlying conservative (i.e., undamped and unforced) system, the approach relies on three fundamental properties:

1. Forced responses of nonlinear systems at resonance occur in the neighborhood of NNMs.
2. According to the invariance property, motions that start out in the NNM manifold remain in it for all time.
3. For weak to moderate damping, its effect on the transient dynamics may be considered to be purely parasitic. The damped invariant manifold can therefore be approximated by the undamped invariant manifold. The free damped dynamics closely follows the NNM of the underlying undamped system.

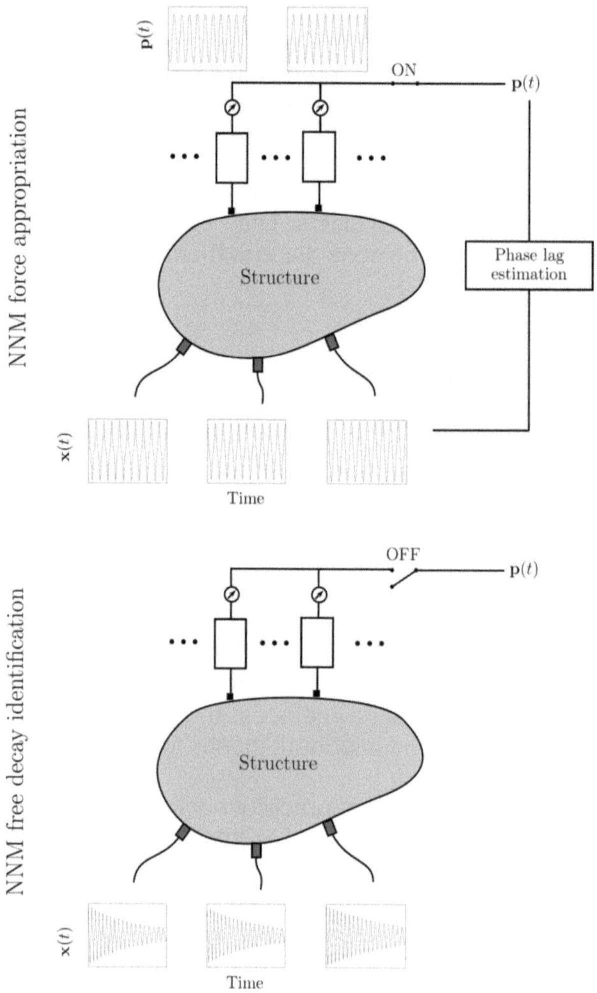


Figure 3: Proposed methodology for experimental modal analysis of nonlinear systems.

From a practical viewpoint, the overall procedure for nonlinear EMA may be viewed as forced vibration testing where the appropriate force is applied as a burst excitation through several exciters for inducing a single-NNM decay response.

### 3.1 NNM force appropriation

For linear systems, force appropriation is usually performed by adapting the frequency and the amplitude distribution of the multi-point excitation. According to the phase lag quadrature criterion (also called phase resonance criterion), a linear damped structure vibrates according to one of the normal modes of the underlying conservative system if all degrees of freedom vibrate synchronously with a phase lag of 90 degrees with respect to the harmonic excitation. Hence, if the phase quadrature criterion is verified during the experimental testing, a single undamped normal mode is isolated, and the natural frequency and the mode shape can be identified.

The NNM framework is used herein to extend force appropriation to nonlinear structures in order to isolate and extract a single NNM at a time. To this end, the forced response of a nonlinear structure with linear viscous damping is considered

$$\mathbf{M}\ddot{\mathbf{x}}(t) + \mathbf{C}\dot{\mathbf{x}}(t) + \mathbf{K}\mathbf{x}(t) + \mathbf{f}_{NL}(\mathbf{x}(t)) = \mathbf{p}(t) \tag{1}$$

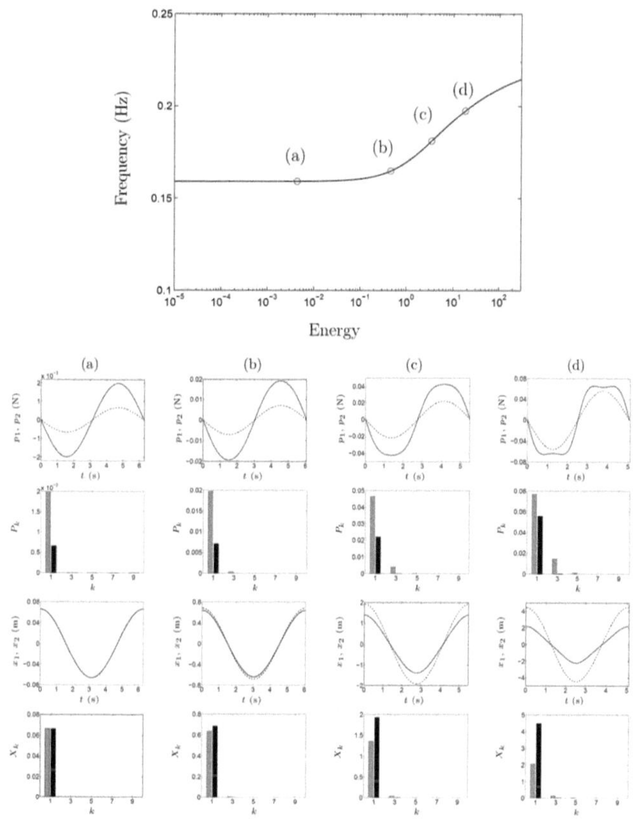


Figure 4: Frequency-energy dependence of the appropriate excitation for the 2DOF system. Top plot: FEP of the in-phase NNM of the underlying undamped system. Bottom plots (from top to bottom): time series of the appropriate excitation (------:$p_1(t)$; - - -: $p_2(t)$); Fourier coefficients of the appropriate excitation (grey: $p_1$; black: $p_2$); time series of the corresponding NNM motion (------: $x_1(t)$; - - -: $x_2(t)$); Fourier coefficients of the NNM motion (grey: $x_1$; black: $x_2$).

Extracting a given NNM motion of the underlying conservative system by means of appropriate excitation is equivalent to assume $\mathbf{x}(t) = \mathbf{x}_{NNM}(t)$ in the equations of motion. Because an undamped NNM motion is defined as a periodic solution of the underlying conservative system, the excitation vector achieving a perfect appropriation of the damped system is given by

$$\mathbf{p}_{NNM}(t) = \mathbf{C}\dot{\mathbf{x}}_{NNM}(t) \tag{2}$$

This relationship shows that the appropriate excitation is periodic and has the same frequency components as the corresponding NNM motion (i.e., generally including multi-harmonic components).

An NNM motion is now expressed as a Fourier cosine series

$$\mathbf{x}_{NNM}(t) = \sum_{k=1}^{\infty} \mathbf{X}_k^{NNM} \cos k\omega t \tag{3}$$

Where $\omega$ is the fundamental pulsation of the NNM motion and $\mathbf{X}_k$ is the amplitude vector of the $k$th harmonic. In this paper, this type of motion is referred to as monophase NNM motion due to the fact that the displacements of all DOFs reach their extreme values simultaneously. The expression of the corresponding appropriate excitation is given by

$$\mathbf{p}_{NNM}(t) = -\sum_{k=1}^{\infty} \mathbf{C}\mathbf{X}_k^{NNM} k\omega \sin k\omega t \tag{4}$$

Comparing equations (3) and (4), the excitation of a monophase NNM is thus characterized by a phase lag of 90 degrees of each harmonics with respect to the displacement response.

One important feature of the appropriate excitation is that it is energy-dependent. In particular, it is characterized by the same frequency-energy dependence as the corresponding NNM. This is illustrated for the 2DOF example in Figure 4 where the FEP of the in-phase NNM is depicted. The appropriate excitation and the resulting NNM motion are depicted for four different energy levels. Their Fourier coefficients are also represented. Clearly, when progressing from low to high energies, the fundamental frequency of the appropriate excitation increases, which is due to the hardening behavior of the system. In addition, multiple harmonics are necessary to induce single-NNM behavior.

### 3.2 Phase lag quadrature criterion for NNM force appropriation

The previous section has demonstrated that nonlinear systems can successfully be forced according to a given NNM at a specific energy level through force appropriation. An indicator highlighting that the NNM appropriation has effectively been achieved would be particularly useful. To this end, we generalize the phase lag quadrature criterion (or phase resonance criterion) to nonlinear systems.

The forced response of the damped system (1) is examined when it vibrates according to a monophase periodic motion with a phase lag of 90 degrees of each harmonics with respect to the periodic excitation, i.e.,

$$\mathbf{x}(t) = \sum_{k=1}^{\infty} \mathbf{X}_k \cos k\omega t, \;\; \mathbf{p}(t) = \sum_{k=1}^{\infty} \mathbf{P}_k \sin k\omega t \tag{5}$$

As the response is even with respect to the time, the nonlinear restoring force can be written as a cosine series

$$\mathbf{f}_{NL}\left[\mathbf{x}(t)\right] = \sum_{k=1}^{\infty} \mathbf{F}_{NL,k} \cos k\omega t \tag{6}$$

We then obtain

$$-k^2\omega^2\mathbf{M}\sum_{k=1}^{\infty} \mathbf{X}_k \cos k\omega t - k\omega\mathbf{C}\sum_{k=1}^{\infty} \mathbf{X}_k \sin k\omega t + \mathbf{K}\sum_{k=1}^{\infty} \mathbf{X}_k \cos k\omega t + \sum_{k=1}^{\infty} \mathbf{F}_{NL,k} \cos k\omega t = \sum_{k=1}^{\infty} \mathbf{P}_k \sin k\omega t \tag{7}$$

By balancing the coefficients of respective harmonics, it follows that

$$\begin{aligned} -k^2\omega^2\mathbf{M}\mathbf{X}_k + \mathbf{K}\mathbf{X}_k + \mathbf{F}_{NL,k} &= 0 \\ \mathbf{P}_k &= -k\omega\mathbf{C}\mathbf{X}_k \end{aligned} \tag{8}$$

So, the external force is given by

$$\begin{aligned} \mathbf{p}(t) &= \sum_{k=1}^{\infty} \mathbf{P}_k \sin k\omega t \\ &= -\sum_{k=1}^{\infty} k\omega\mathbf{C}\mathbf{X}_k \sin k\omega t \\ &= \mathbf{C}\dot{\mathbf{x}}(t) \end{aligned} \tag{9}$$

and the periodic response x(t) is a solution of the underlying conservative system. The response x(t) is therefore an undamped NNM motion of the system: the NNM force appropriation is realized.

In conclusion, **the phase lag quadrature criterion, valid for linear systems, can be generalized to monophase NNM motions of nonlinear structures, where the phase lag is defined with respect to each harmonics of the monophase signals**.

In other words, if the response (in terms of displacements or accelerations) across the structure is a monophase periodic motion in quadrature with the excitation, the structure vibrates according to a single NNM of the underlying conservative system. The phase lag of the generated monophase excitation with respect to the response can thus be used as an indicator of the NNM appropriation.

No direct constructive method exists to determine the appropriate excitation of a given NNM. Such an excitation has to be derived through successive approximations based on this indicator. For nonlinear structures, in addition to the spatial distribution of the multi-point excitation, the amplitude distribution of harmonic terms has also to be tuned.

The phase lag quadrature criterion is now used for estimating the quality of the NNM appropriation. The periodic forced responses of the damped version of the 2DOF system in Figure 1 to a harmonic force of frequency $\omega$ and amplitude $F$ applied to the first DOF are analyzed. An imperfect force appropriation (i.e., $p_2(t)=0$ and purely harmonic excitation) is purposely considered to investigate the robustness of the proposed procedure.

The nonlinear frequency responses close to the resonance of the in-phase mode were computed using shooting and continuation methods and are depicted in Figure 5. They are given in terms of amplitude and phase lag (of the fundamental frequency component) of the displacement response for increasing forcing amplitudes. It is observed that the phase quadrature criterion is almost verified close to the forced resonance. For $F$=0.2N, i.e., at the point marked by a square, the phase lag is equal to 90 degrees and 91 degrees for the first and second DOFs, respectively. Figure 6 represents the time series of the displacement response for $F$=0.2N. Clearly, the displacement is practically monophase with a phase lag around 90 degrees with respect to the excitation $p_1(t)$.

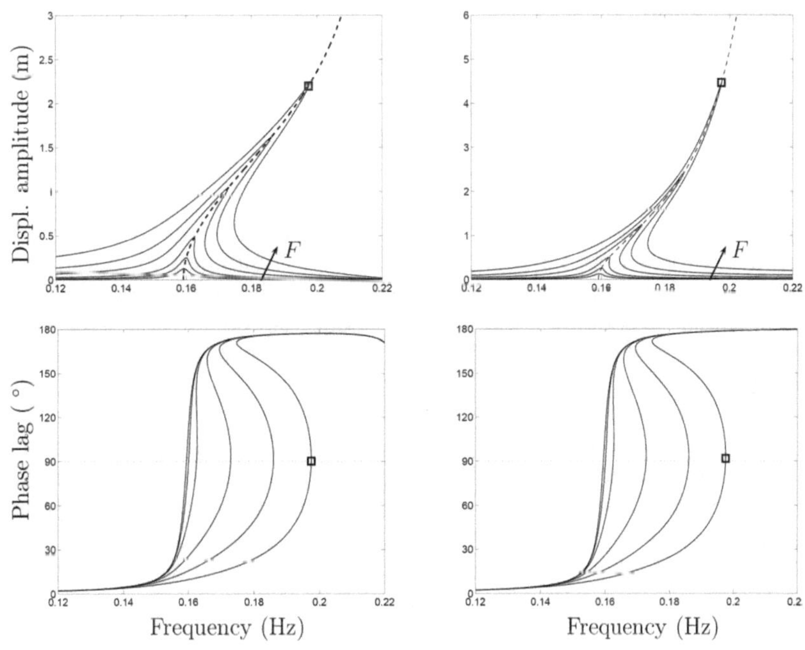


Figure 5: Nonlinear frequency responses of the damped 2DOF system close to the first resonant frequency (6 different forcing amplitudes $F$: 0.005N, 0.01N, 0.02N, 0.05N, 0.1N, 0.2N). The dashed line is the backbone of the first undamped NNM (computed by means of the numerical algorithm). Top plots: displacement amplitude. Bottom plots: phase lag of the displacement with respect to the excitation. Left plots: $x_1$; right plots: $x_2$.

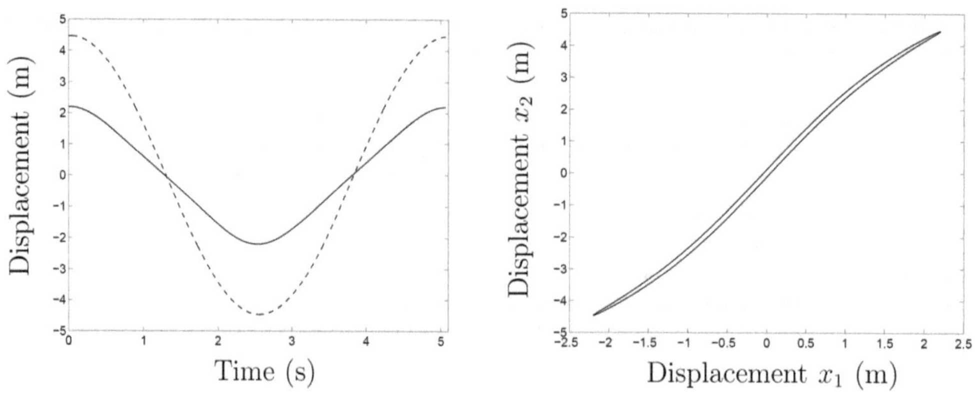


Figure 6: Forced response (corresponding to $F$=0.2N and marked by a square in Figure 5) of the damped 2DOF system Left plot: time series. (------: $x_1(t)$; - - -: $x_2(t)$). Right plot: motion in the configuration space.

These results also confirm that forced responses of nonlinear systems at resonance occur in the neighborhood of NNMs. The backbone of the in-phase undamped NNM is expressed in terms of amplitude and is displayed using a dashed line in Figure 5. This backbone curve traces the locus of the frequency response peaks.

It is interesting that no forcing on the second DOF and no higher harmonic terms were necessary to isolate this high-energy NNM motion. This is an appealing feature for future practical realizations, at least for structures with relatively well-separated modes. A constructive method for inducing single-NNM behavior could be to perform successive adjustments of a stepped sine excitation until the phase lag criterion is verified.

### 3.3 NNM free decay identification

By means of nonlinear force appropriation, the forced response of a damped system may be restricted to a single undamped NNM at a specific energy level. In view of the frequency-energy dependence, successive appropriate excitations at different force levels have to be considered to provide a complete characterization of the NNM of interest. This may complicate the experimental realization.

The alternative strategy proposed here consists in exploiting the NNM invariance property. When a high-energy NNM motion is isolated using force appropriation, the excitation is stopped to obtain the resulting free damped response. Due to invariance, this free decay response initiated on the undamped NNM remains close to it when energy decreases. Using this procedure, the energy dependence of the NNM modal curves and the corresponding frequencies of oscillation may easily be extracted from the single-mode free damped response at each measurement locations.

As mentioned above, according to invariance, the resulting free damped response closely follows the corresponding undamped NNM provided that the damping is moderate. The relation between the two responses is only phenomenological, nevertheless it enables one to interpret the damped response in terms of NNM motions of the underlying undamped system. In fact, the damped manifold corresponds to the exact invariant manifold of the damped dynamics. However, for lightly damped structures, the latter may be approximated by the undamped NNM that can be viewed as an attractor of the free damped response.

For illustration, the 2DOF system is considered. From an appropriated in-phase NNM motion, the resulting free damped response when the excitation is removed is depicted in Figure 7. Figure 8 compares the manifold corresponding to the in-phase undamped NNM with the free damped response represented in the same projection of phase space. Clearly, it confirms that the free damped response traces the NNM manifold of the underlying undamped system with very good accuracy when energy decreases.

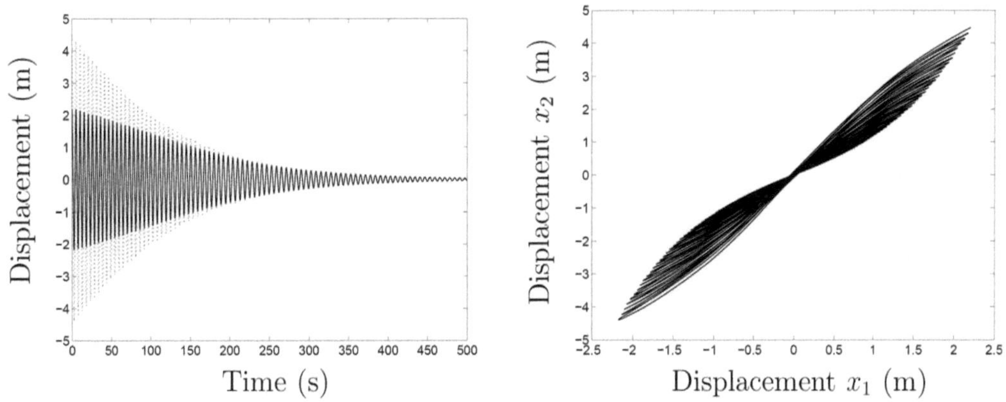


Figure 7: Free response of the damped 2DOF system initiated from the in-phase NNM motion. Left plot: time series (------: $x_1(t)$; - - -: $x_2(t)$). Right plot: motion in the configuration space.

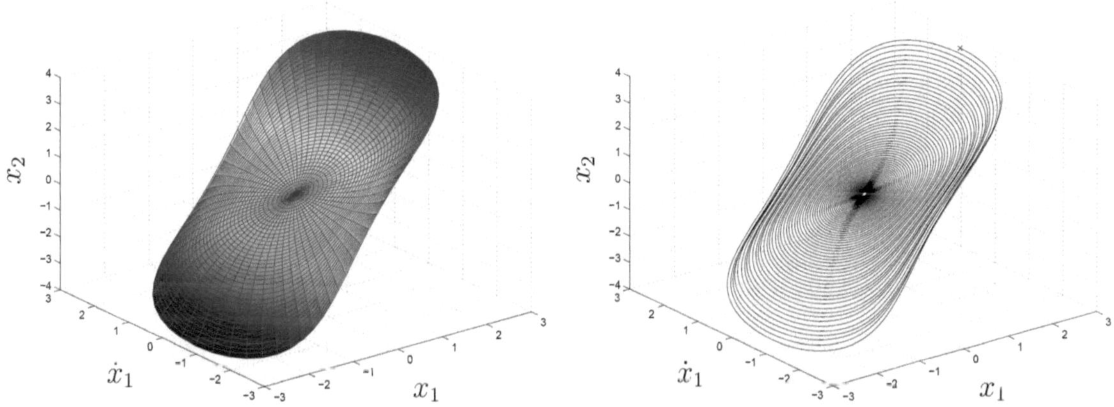


Figure 8: Invariant manifold of the in-phase NNM of the 2DOF system. Left plot: NNM manifold of the underlying undamped system. Right plot: free response of the damped system

**NNM Extraction**

As a result, the NNM modal curves may be extracted directly from the single-NNM free decay response. They are obtained by representing the time series in the configuration space for one oscillation around specific time instants, associated with different energy levels.

To compute the oscillation frequency of NNMs, time-frequency analysis is considered. Time-frequency analysis is a versatile tool for analyzing nonstationary signals; i.e., signals whose spectral contents vary with time. It has been successfully exploited in structural dynamics, e.g., for linear and nonlinear system identification. The continuous wavelet transform (CWT) is used in this paper. In contrast to the Fourier transform, which assumes signal stationarity, the CWT involves a windowing technique with variable-sized regions. Small time intervals are considered for high-frequency components, whereas the size of the interval is increased for lower-frequency components. The CWT can therefore track the temporal evolution of the instantaneous frequencies, which makes it an effective tool for analyzing nonlinear signals. Using the CWT, the oscillation frequency of the NNM may then be extracted from the time series of the free damped response. The usual representation of the transform is to plot its modulus as a function of time and frequency in a three-dimensional or contour plot. For illustration, the CWT of the free decay response of the 2DOF system represented in Figure 7 is shown in Figure 9. The instantaneous frequency decreases with time, and hence with energy, which reveals the hardening characteristic of the system.

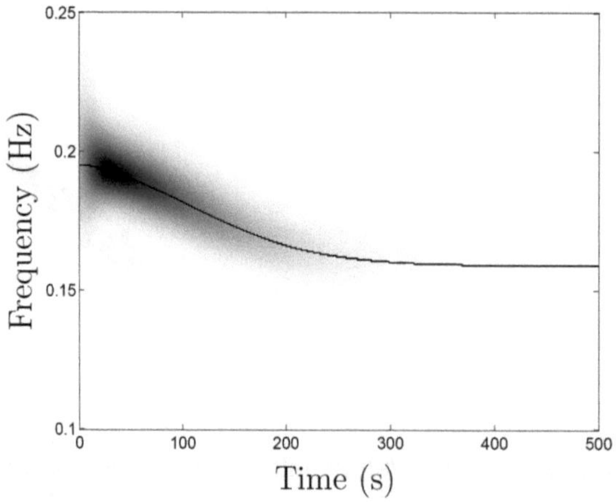


Figure 9: Extraction of oscillation frequencies of the NNM during the free decay. Temporal evolution of the instantaneous frequency of the single-NNM free response of the damped 2DOF system computed using the CWT. The solid line is the ridge of the transform; i.e., the locus of the maxima at each time instant.

**Reconstructed FEP**

When the total energy (i.e., the sum of the kinetic and potential energies) in the system can be determined, the experimental FEP can be reconstructed in a straightforward manner by substituting the instantaneous energy in the system for time:
(i) the backbone expressing the frequency-energy dependence of the NNM is provided by the CWT,
(ii) the obtained modal curves around different energy levels are superposed in the plot.

For the 2DOF example, the experimental FEP calculated from the time series of the free damped response is represented in Figure 10. It displays the experimental backbone determined through the CWT and the experimental modal curves. For comparison, the theoretical FEP of the undamped NNM computed in Section 2 from the equations of motion is also illustrated. Except the CWT edge effects, a perfect agreement is obtained between the two FEPs, which shows again that the undamped NNM is an attractor for the damped trajectories. In the present case, the linear modal damping ratios are 1% and 0.6%, but we note that this result holds for higher damping ratios.

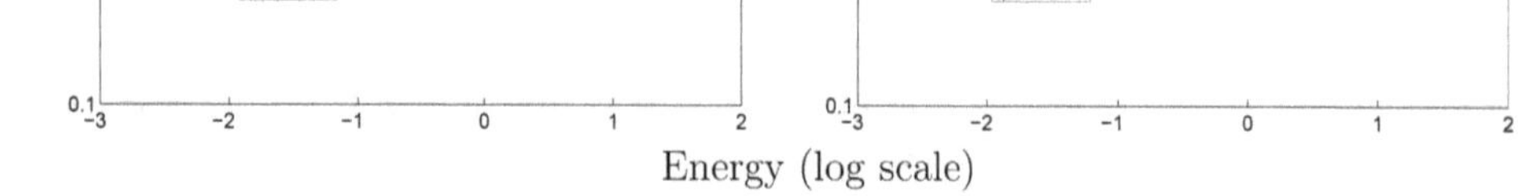


Figure 10: Frequency-energy plot of the in-phase NNM of the 2DOF system. Left plot: theoretical FEP computed by means of the numerical algorithm from the undamped system. Right plot: experimental FEP calculated directly from the time series of the free damped response using the CWT. The solid line is the ridge of the transform.

In summary, it is thus validated that the free damped dynamics can be interpreted based on the topological structure of the NNM of the underlying conservative system. As a result, one can fully reconstruct the FEP and extract the modal curves together with the oscillation frequencies of the NNM using the proposed procedure.

## 4. CONCLUDING REMARKS

Realizing that nonlinearity is a frequent occurrence in engineering structures and that linear EMA is of limited usefulness in this context, the present paper is an attempt to develop nonlinear EMA by targeting the extraction of NNMs from time series.

Because modal superposition is no longer valid, dynamic testing of nonlinear structures is realized through a nonlinear phase resonance method, which relies on the extension of the phase lag quadrature criterion. Specifically, if the forced response across the structure is a monophase periodic motion in quadrature with the excitation, a NNM vibrates in isolation. Once the NNM appropriation is achieved, the complete frequency-energy dependence of that nonlinear mode can be identified during the free decay response using time-frequency analysis. Eventually, an experimental FEP for one specific NNM can be obtained, and the procedure can be applied for all NNMs of interest. To relate the NNMs of the underlying undamped system to those extracted from the experimental data, the procedure assumes moderately damped systems possessing elastic nonlinearities.

This two-step methodology paves the way for a practical nonlinear analog of EMA, which may be applied to strongly nonlinear systems. For instance, it can certainly be a solid basis for extending standard ground vibration testing to nonlinear aircrafts. Through the combination of EMA with TMA, finite element model updating and validation of nonlinear structures is also within reach.

Further research will first deal with the experimental demonstration of these theoretical findings using an existing nonlinear beam. Because nonlinear systems undergo bifurcations, the robustness of the procedure will also be carefully assessed against NNM modal interactions and shrinking basins of attractions.

## 5. REFERENCES


1. K. Carney, I. Yunis, K. Smith, C.Y. Peng, *Nonlinear dynamic behavior in the Cassini spacecraft modal survey*, Proceedings of the International Modal Analysis Conference (IMAC), Orlando, USA (1997).
2. N. Okuizumi, M.C. Natori, *Nonlinear vibrations of a satellite truss structure with gaps*, Proceedings of the 45th AIAA Structures, Structural Dynamics & Materials Conference, Palm Springs, USA (2004).
3. G. Kerschen, K. Worden, A.F. Wakakis, J.C. Golinval, *Past, present and future of non-linear system identification in structural dynamics*, Mechanical Systems and Signal Processing 20 505-592 (2006).
4. M.F. Platten, J.R. Wright, G. Dimitriadis, J.E. Cooper, *Identification of multi-degree of freedom non-linear systems using an extended modal space model*, Mechanical Systems and Signal Processing 23 8-29 (2009)
5. D. Goge, U. Fullekrug, M. Sinapius, M. Link, L. Gaul, Advanced test strategy for identification and characterization of nonlinearities of aerospace structures, AIAA Journal 43 974-986 (2005).
6. G. Kerschen, M. Peeters, J.C. Golinval, A.F. Vakakis, *Nonlinear normal modes, Part I: A useful framework for the structural dynamicist*, Mechanical Systems and Signal Processing 23 170-194 (2009).
7. M. Peeters, R. Viguie, G. Serandour, G. Kerschen, J.C. Golinval, *Nonlinear normal modes, Part II: Toward a practical computation using numerical continuation techniques*, Mechanical Systems and Signal Processing 23 195-216 (2009).
8. F. Georgiades, M. Peeters, G. Kerschen, J.C. Golinval, M. Ruzzene *Modal analysis of a nonlinear periodic structure with cyclic symmetry*, AIAA Journal 47 1014-1025 (2009).

Proceedings of the IMAC-XXVIII
February 1–4, 2010, Jacksonville, Florida USA


# Damage Detection of Reinforced Concrete Structures Using Nonlinear Indicator Functions

Chin-Hsiung Loh
Professor
National Taiwan University
Civil Eng. Department
Taipei 10617, TAIWAN
e-mail: lohc0220@ccms.ntu.edu.tw

JIan-Hong Mao
Former graduate student
National Taiwan University
Civil Eng. Department
Taipei 10617, TAIWAN
e-mail:r96521209@ntu.edu.tw

Jhih-Ren Huang
Former graduate student
National Taiwan University
Civil Eng. Department
Taipei 10617, TAIWAN
e-mail:r96521207@ntu.edu.tw

## ABSTRACT

In this study, signal processing approaches and nonlinear indicators are used to measure seismic responses of reinforced concrete structures. To analyze structural nonlinearity, an equivalent linear system with time-varying model parameters, singular spectrum analysis to elucidate residual deformation, and wavelet analysis to yield the energy distribution among components are adopted to detect the nonlinearity. Then, damage feature extraction is conducted using both the Holder exponent and the Level-1 detail of the discrete wavelet component. Finally, the modified Bouc-Wen hysteretic model and the system identification process are employed to shaking table test data of the RC frames to evaluate the physical parameters, including the stiffness degradation, the strength deterioration and the pinching hysteresis. The stiffness and strength degradation of the RC frame in relating to the degree of ground shaking and damage index are discussed.

## INTRODUCTION

Building structures typically exhibit nonlinear and inelastic behavior under severe dynamic loading, such as during earthquakes and in severe winds. Generally, the maximum inter-story drift has been shown to be a good measure of the extent of nonstructural damage. This measure neglects the effect of cumulative damage. The effect of accumulated damage to reinforced concrete members and structures under a seismic load is even more important than the maximum inter-story drift. Inelastic behavior commonly manifests as a hysteresis loop under cyclic loading. Most important, the performance and safety of reinforced concrete structures under severe earthquake loading are determined by nonlinear hysteresis. In any assessment of the damage to a reinforced concrete structure, nonlinear hysteretic behavior of each member must be considered. The Bouc-Wen model of smooth hysteresis [1] is receiving increasing attention as a means of specifying analytically a range of shapes of hysteretic cycles. Ikhouane *et al.* characterized various classes of Bouc-Wen model with reference to their bounded input-bounded output stability *et al.*[2].

Unlike a theoretical model, a damage identification system that is based on vibration measurements is an effective means of detecting damage to whole structures. Vibration-based damage detection approaches assume that changes in a structure's integrity influence the measured vibration signals, enabling damage to be detected independently of the modal parameters. One important issue in global vibration-based damage assessment approaches concerns the choice of damage indices. Doebling *et al.* [3] summarized the historic development of damage assessment approaches. The monitoring of the condition of dynamic systems based on vibration signatures is conducted using the discrete wavelet transform (DWT), which may be regarded as an extension of the traditional Fourier Transform with an adjustable window location and size. The wavelet packet transform (WPT) is one extension of the DWT that provides a complete level-by-level decomposition. Yen *et al.* [4] adopted wavelet packet feature extraction to monitor vibrations. Sun *et al.* [5] employed WPT with redundant basis functions to provide an arbitrary time-frequency resolution in the detection of damage. Han *et al.* investigated the application of wavelet packet sifting for detecting damage to a beam structure [6]. Using non-linear features is believed to improve the damage classification capacity of a damage detection scheme.

In this study, nonlinear system identification and feature extraction processes are used to investigate the physical parameters in the degrading hysteretic model of a reinforced concrete frame and the severity of damage thereto. Six reinforced concrete frame test data collected in response to various degrees of seismic excitation, including seismic response data, ambient test data and cyclic loading test data, are used to study the severity of damage. The proposed methods identify physical parameters such as stiffness degradation, strength deterioration and the pinching behavior of the reinforced concrete structure, as well as the feature extraction using nonlinear indices to determine the severity of damage.

## SHAKING TABLE TESTS OF ONE-STORY TWO- BAY RC FRAMES

Six reinforced single-story two-bay reinforced concrete frames were constructed and designed with the same dimensions and the design details. *Figure 1* presents the test frame. The height of each frame structure is 2.0 m and the length of each span is 2.0 m. The designed compressive strength of the concrete is 210kg/cm$^2$ and the designed strength of the steel is 4200 kg/cm$^2$. The frame comprised a T-beam, with a floor dimension of the first story of 4.7 m × 0.7 m, and three columns with cross-sectional dimensions of 20 cm × 20 cm. The total mass of each frame is 8490 kg (which includes 4000 kg of lead ballast). *Table 1* presents in detail the dimensions of the test structure. Both a shaking table test and a cyclic loading test are performed on the structure following the approaches for each frame that are presented in *Table 2*. Accelerometers and LVDT are installed in the basement and on top of the floor. Damage is caused to each frame in the shaking table test in which various intensities of base input excitations are used. Chi-Chi earthquake ground motion data from station TCU082 (NS-direction with duration of 98 sec) are adopted as an input excitation. Damage is determined from seismic response data for the following four specimens; Specimen RCF6 (with input PGA=649gal), Specimen RCF2 (with input PGA=840gal), Specimen RCF4 (with input PGA=1186gal) and Specimen RCF3 (with input PGA=1310gal).

The maximum inter-story drift ratio and the peak floor acceleration can be determined directly from the response measurements of these frame structures made during the shaking table tests, as presented in *Table 3*. For the RCF3 specimen, the peak floor acceleration achieved using the shaking table was 1287 gal, which induced a maximum inter-story drift ratio of the structure of the structure of 4.46%, while for the RCF6 specimen, it induced a maximum inter-story drift ratio of 1.35%. *Table 3* clearly indicates that these four specimens experienced different degrees of damage, and provides important information for determining the different degrees of damage to the RC frame in response to various degrees of ground excitation. *Figure 2* plots the load-displacement hysteresis of these four specimens based on the displacement and acceleration measurements made in the shaking table tests. Hysteresis is important in determining the nonlinear response of reinforced concrete structures as well as the damage. Variations in the load-displacement relationship among these four tests are obvious. The following section discusses the stiffness and strength degradation of these specimens.

## SIGNAL-BASED NONLINEAR IDENTIFICATION

The observed load-displacement hysteresis curves and the maximum inter-story drift ratios of the four test specimens clearly show the structural nonlinearity that is caused by the decrease in stiffness of the structures, and the consequent change in their dynamic responses (inter-story drift). To detect damage to the RC frame, signal-based identification can be conducted. Besides, the benefit of the nonlinear effect is that it amplifies the damage-sensitive features of the measured data. Therefore, damage-related features can also be identified. In this section, several nonlinear indices are extracted from the response measurements. The degree of nonlinearity that is induced in these specimens by various levels of base excitation will be identified.

Table 1: Dimensions of the test reinforced concrete structure

| | |
|---|---|
| Floor Height | 2.0 m |
| Span Length | 2.0 m |
| Center Column | 20 cm x 20 cm ((8 #4 steel bars) |
| Side Column | 20cm x 20cm (4 #4 steel bars) |
| Beam | 20 cm x 10 cm<br>(Top 3 #3 steel bars; bottom 2 #3 steel bars) |
| Transverse Steel | #3 with 10cm spacing |
| Total Weight | 8590 kg |
| Floor Weight | 6454kg (include 4000 kg lead ballistic) |

Table 2: Test procedure on six reinforced concrete frames.

| Test Procedure | Ambient Test (before shaking table test) | Shaking Table Test (Input Level) | Cyclic Loading Test | Ambient Test (after shaking table test |
|---|---|---|---|---|
| RCF1 | Acceleration Velocity | None | 5% Drift Ratio | Acceleration Velocity |
| RCF2 | Acceleration Velocity | 840gal | 5% Drift Ratio | Acceleration Velocity |
| RCF3 | Acceleration Velocity | 1310gal | 5% Drift Ratio | Acceleration Velocity |
| RCF4 | Acceleration Velocity | (a) 1186gal<br>(b) 668gal | 5% Drift Ratio | Acceleration Velocity |
| RCF5 (with additional dead load) | Acceleration Velocity | 1249gal | 5% Drift Ratio | Acceleration Velocity |
| RCF6 | Acceleration Velocity | 649gal | 5% Drift Ratio | Acceleration Velocity |

Table 3: Input peak ground acceleration, maximum acceleration response and maximum inter-story drift of the four test specimens.

| Specimen No. | Excitation | PGA(gal) | Max. Absolute Acceleration (gal) | Max. Inter-Story Drift Ratio (%) |
|---|---|---|---|---|
| | White Noise | 29 | 24 | 0.06 |
| RCF6 | TCU082 | 622 | **750** | **1.35** |
| | White Noise | 30 | 15 | 0.04 |
| | White Noise | 27 | 41 | 0.02 |
| RCF2 | TCU082 | 815 | **1241** | **2.43** |
| | White Noise | 30 | 28 | 0.05 |
| | White Noise | 29 | 46 | 0.03 |
| RCF4 | TCU082 | 1157 | **1305** | **3.29** |
| | White Noise | 32 | 25 | 0.06 |
| | White Noise | 24 | 40 | 0.04 |
| RCF3 | TCU082 | 1287 | **1216** | **4.46** |
| | White Noise | 35 | 25 | 0.06 |

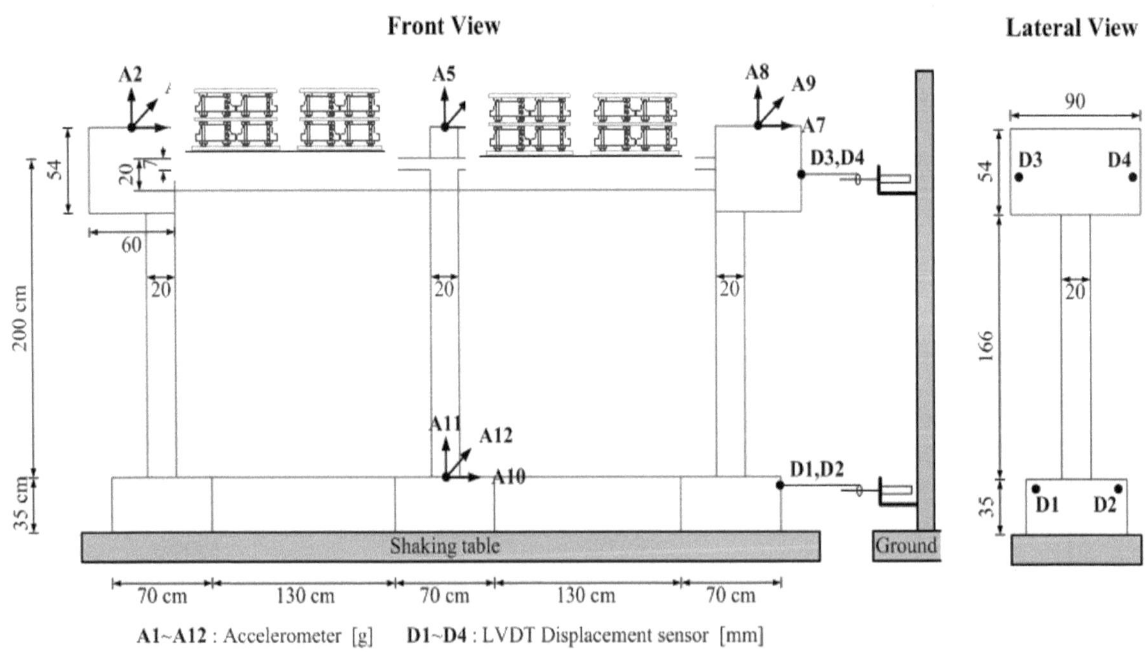


Figure 1: Front view and side view of the test specimen. The locations of accelerometers and the LVDT are also identified in this figure.

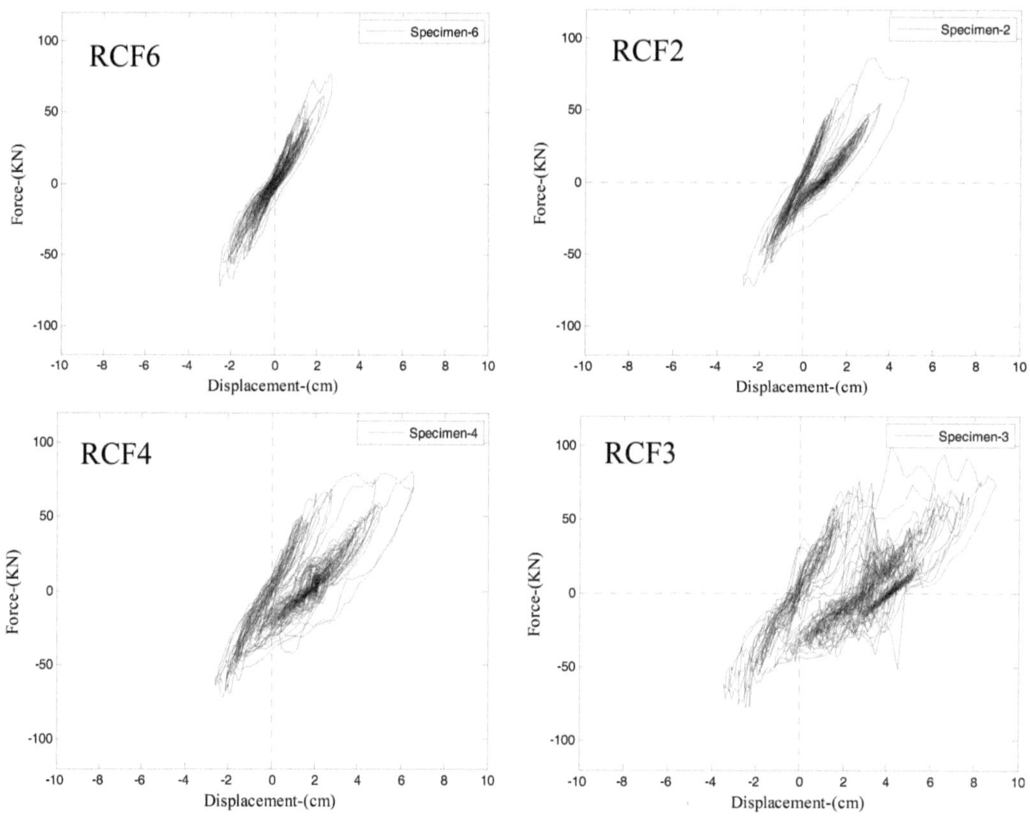


Figure 2: Plot of the inter-story load-displacement hysteretic diagram of the frame structure (Four hysteresis loops from four different test specimens).

***Equivalent linear system with time-varying model parameters:*** One of the intuitively nonlinear indices for studying nonlinearity is to study the time-varying parameters of the model of the system by considering its equivalent linear system using a moving time window. Observing the time-varying natural frequency of the system and damping ratio of the equivalent linear single-degree-of-freedom dynamic system yields the nonlinearity of the system. *Figure 3* plots the fundamental natural frequency of the identified time-varying system. The variations in the natural frequency with time and the level of excitation are obvious.

***Residual Deformation Analysis:*** Material nonlinearities that are associated with excessive deformation which can cause a structure to behave in a nonlinear manner, and yielding is accompanied by permanent deformation. Singular spectrum analysis (SSA) is applied to the recorded displacement data of the specimen under an earthquake excitation to estimate the residual and the permanent deformations of the specimen from the seismic response data of the frame structure due to yielding [8]. SSA is employed as an alternative to traditional digital filtering. It is a novel non-parametric approach that is based on the principles of multivariate statistics. The original time series is decomposed into various additive time series, each of which can be easily identified as being part of either the modulated signal or the random noise. The method begins by generating a Hankel matrix from the original time series by sliding a window that is shorter than the original series. This matrix is decomposed by SVD into many elementary matrices with decreasing norm. *Figure 4* presents the estimated time-dependent residual deformation of the four tested frames (RCF6, RCF2, RCF4 and RCF3). Significant residual deformation is observed under strong ground excitation such as exhibited by the RCF3 specimen.

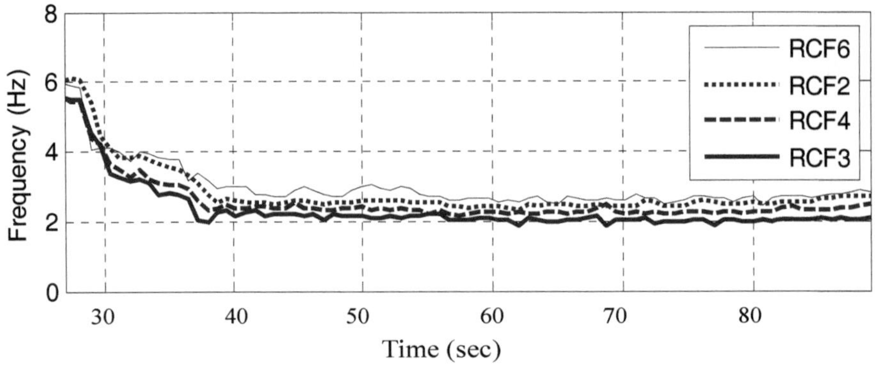


Figure 3: Identified time-varying model frequency of the four frames using equivalent linear model.

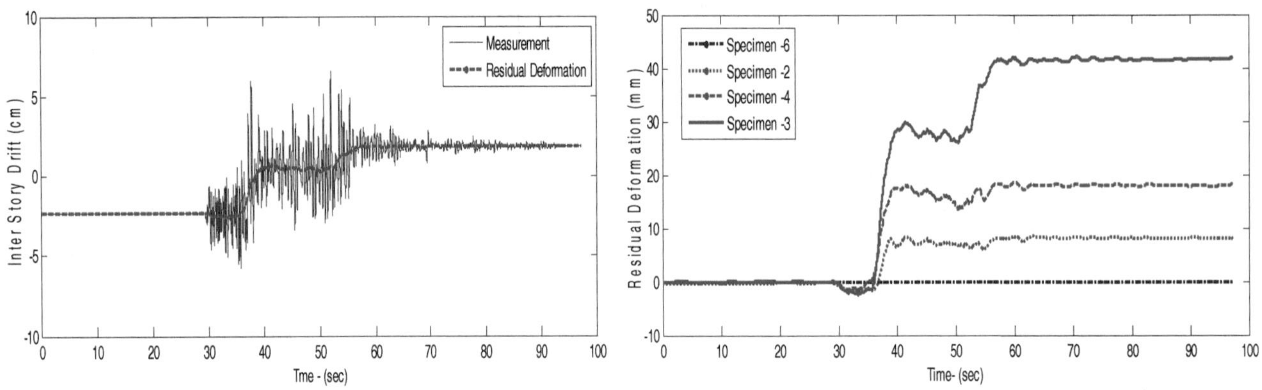


Figure 4: (a) Estimated residual deformation on top of the recorded Inter-story drift of specimen RCF4, (b) Identified residual deformation of the four specimens (RCF6, RCF2, RCF4, and RCF3).

***Distribution of Component Energy using Wavelet Packet Transform**:* Since using the wavelet packet analysis of the signal can be regarded as a sifting bank with adjustable time and frequency resolution, a sifting process that is based on wavelet packet decomposition to analyze a non-stationary signal can be formulated and used for monitoring structural health. The Wavelet Packet Transform analysis is similar to Discrete Wavelet Transform, but with the only difference that, in addition to decomposing the wavelet approximation component at each level, a wavelet detail component is also decomposed to yields *its* own approximation and detail component. Wavelet packet decomposition is performed up to the j-th level. Then, the structural response is decomposed using wavelet packet analysis (using bior6.8) and the central frequency of the *i-th* component function is obtained. In the j-th level decomposition, the distribution of component energy of each decomposed signal can be generated with respect to its central frequency. The distribution of the wavelet packet component energy of the decomposed response data can be taken as the distribution of energy that is stored in the frequency band. The distribution of component energy obtained from the acceleration response data of the four specimens can be used not only to extract the dominant frequency component of the response signal but also to determine the distribution of component energy in the frequency domain. Based on the wavelet decomposition of the acceleration responses of these four specimens, *Figure 5* plots the percentage distribution of wavelet component energy. According to the date, the distribution of the dominant energy of the wavelet components from the specimen subjected to a stronger excitation level is shifted to lower frequency.

## IDENTIFICATION OF DAMAGE BY FEATURE EXTRACTIONS

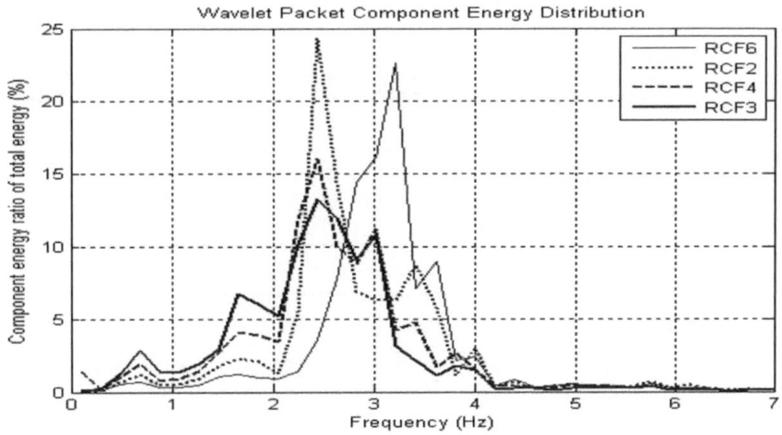


Figure 5: Distribution of component energy of the acceleration response of the four specimens.

To evaluate the severity of damage to a reinforced concrete structure, the damage index of Park and Ang is used. The Park and Ang damage index $DI_g$ is defined as [9, 10],

$$DI_g = \frac{\delta}{\delta_u} + \beta \frac{E_H}{\delta_u F_y} \tag{1}$$

where $\delta_u$ is the ultimate displacement; $F_y$ is the yield force; $E_H$ is the hysteretic energy of the structural element and $\beta$ is a parameter. Both $\delta_u$ and $F_y$ are estimated by performing cyclic loading test of specimen RCF1 (the reference specimen). From the envelope of the cyclic loading curve of specimen RCF1, the maximum yield base shear force $F_y$ is estimated and the ultimate displacement $\delta_u$ , which corresponds to $0.8F_y$ , is also determined. *Figure 6* plots the estimated time-varying damage index for the four specimens during shaking table tests. The other measure of damage of a structural system is the inter-story drift ratio. The inter-story drift ratios of these four specimens are calculated using the data from LVDT response measurements, as presented in *Figure 6*. At $t = 30.0$ sec, $t - 37.5$ sec and $t = 52$ sec, considerable increases in the damage index occurred, consistent with a large inter-story drift ratio.

Estimating the damage index and the inter-story drift ratio of the RC frame depends on both input and response measurements (both displacement and acceleration data). To identify changes that are indicative of the onset of nonlinear system response, directly from only the response measurements without the input data, some nonlinear index functions can be employed to identify changes that are indicative of the onset of nonlinear system response, directly from only the response measurements without the input data. The following two feature extraction methods are applied to the floor acceleration response data of the four test structures:

***Holder Exponent*** The Holder Exponent is also known as the Lipschitz exponent, which is an index of the regularity of a signal at a particular instant [11]. Conventionally, the Holder exponent is found by Fourier-transforming the signal, which process yields only the global minimum regularity. Wavelet approach is used to evaluate signal regularity in every time step. In this work, the CWT is applied to the recorded acceleration data of the shaking table test to generate a CWT spectrum, The wavelet function, 'Morlet', which has infinite vanishing moment, is employed and the is set from one to 300 in the CWT approach. At each time step, the log of the wavelet modulus is plotted against the log of the scale over a frequency range of 6.0Hz to 50.0Hz and a Gaussian function was applied to smooth the plot. Finally, the gradient (Holder exponent) of the curve is obtained by first-order polynomial curve fitting.

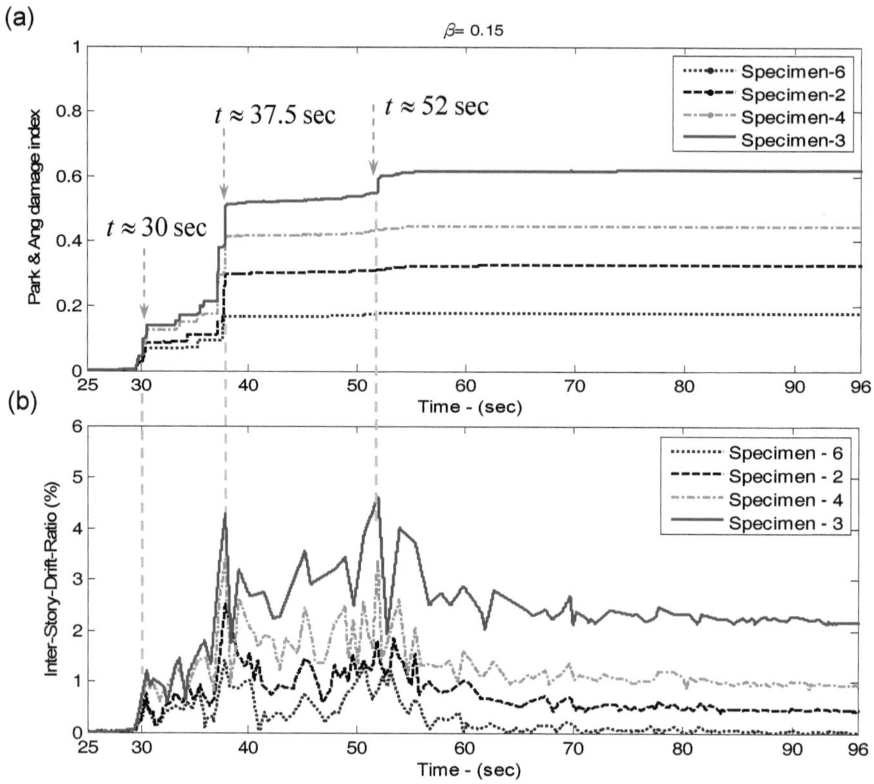


Figure 6 Comparison on the Park & Ang damage index and the inter-story drift ratio with respect to time for the four specimens; (a) Park & Ang damage index, (b) Inter-story drift ratio.

***Level-1 detail component of discrete wavelet transform (DWT)*** DWT can decomposes signals into two components in each decomposition level: the approximate and the detailed. The decomposition is dyadic, and so the detail component contains the upper half frequency band of the original signal. Generally, singularities in the signal are present in very high frequency bands and the Level-1 detail component can reveal such singularities after relatively simple computation. In this work, the level-1 detail component contains information on frequency band from 50 Hz to 100 Hz.

Based on the analysis of the Holder exponent and the details of the DWT, some singularities in the time domain can be identified. These identified singularities are considered in the light of the calculated physical indices, such as the inter-story drift ratio and the Park & Ang damage index. *Figure 7* compares the inter-story drift ratio, the Holder exponent, the level-1 detail component of DWT and the residual component obtained by SSA, using data for specimen RCF6. The estimated singularities in these three nonlinear indices clearly occurred at the times of maximal inter-story drift ratio. For specimen RCF6, as an example, the singularities revealed the various nonlinear indices are all mutually consistent at 30s, 35.0s, 37.5 s, 40 s, and 51.5 s. The singularities identified by the analysis of the detail component of DWT, the Holder exponent and the residual component determine SSA are consistent with significant inter-story drifts. In cases of severe damage, such as evident in specimen RCF4 with a maximum inter-story drift ratio of 3.29%, the singularities revealed by the Holder Exponent, DTW and the residual determined from SSA are more complex because the corresponding structure is severely damaged.

## CONCLUSIONS

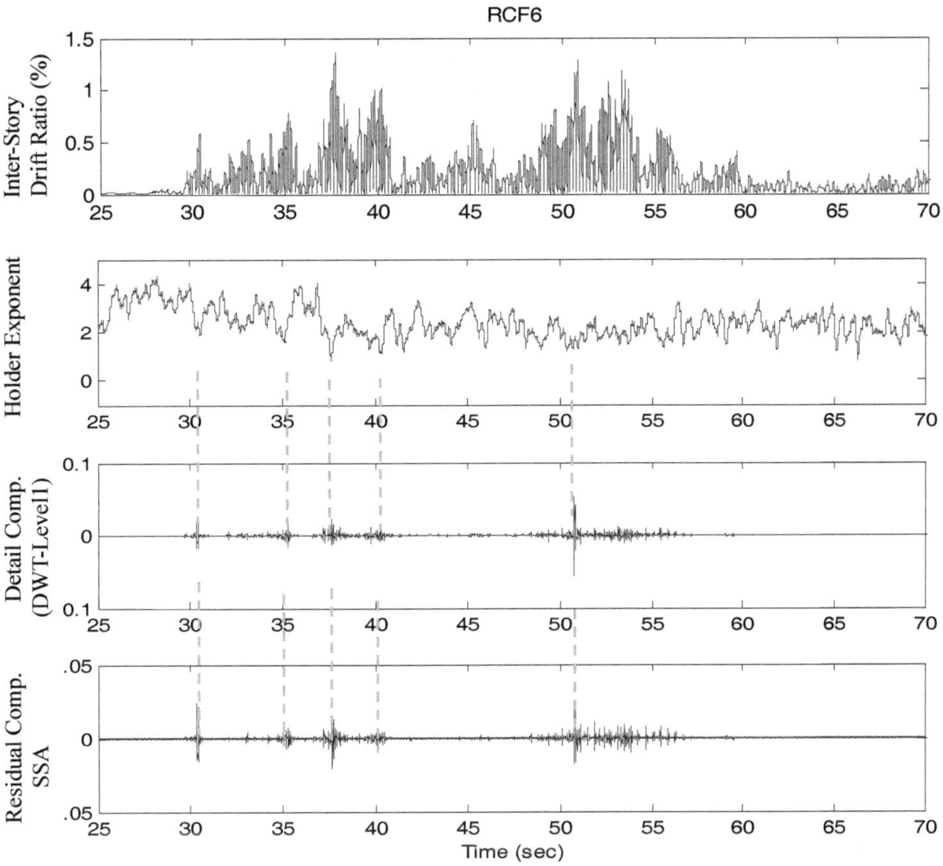


Figure 7: Comparison on Inter-story drift ratio, Holder exponent, Level-1 detail component and residuals estimated from SSA with respect to time from response data of RCF6. Correlation of singularities is identified among different analysis.

The performance and safety of reinforced concrete structures under severe earthquake loading depend on their nonlinear hysteretic behavior. Such a structural system exhibits significant energy dissipation and tends to exhibit a large hysteresis loop with strength and stiffness degradation as well as pinching. In this study, both identification of system nonlinearities and feature extraction are developed to investigate the physical parameters that govern the degradation of hysteretic behavior and the severity of damage of a reinforced concrete structure. First, the nonlinearity of the system under earthquake excitation is studied by examining the time-varying dominant frequency of the equivalent linear system and the component energy distribution of the response. Then, nonlinear indices, such as the Holder exponent and the Level-1 detail component of DWT are calculated to enable the time of significant damage to be determined directly from the response measurements. Singular spectral analysis yields information on Residual deformation. The relationship among the damage index, the inter-story drift ratio, the singularities of the nonlinear indices is examined. A total of six reinforced concrete frames, designed with the same dimensions and the same seismic design code, are individually subjected to different degrees of ground excitation in shaking table tests. Their response data are used to determine the pattern of damage to a reinforced concrete structure with respect to the degree of ground shaking. The seismic response data of these specimens support the following conclusions.

1. The distribution of the wavelet component energy in the response, and the time-varying dominant frequency of the equivalent linear system can be employed to detect the nonlinearity of the system. The permanent deformation of the structure can be estimated by singular spectrum analysis.
2. To determine the time of significant inter-story drift from only acceleration response data, the Holder exponent and Level-1 detail component of DWT can be adopted.
3. To quantify the degree of stiffness degradation, a theoretical model, such as the modified Bouc-Wen hysteretic model with stiffness and strength degradation, must be used, and the system identification approach must be utilized to evaluate the model parameters.
4. The analysis of the shaking table test data for the four RC specimens reveals that under the earthquake excitation, the stiffness of the structure was degraded first, and then the strength deteriorated. Experimental results indicate that the characteristics of the model are consistent with the singularities determined from the nonlinear indices.

## ACKNOWLEDGEMENTS

The authors would like to thank the National Science Council of the Republic of China, Taiwan (Contract No. NSC96-2221-E-002- 121-MY3) and the Research Program of Excellency of National Taiwan University (Contract No. 97R0066-06) for financially supporting of this research. Experimental support from National Center for Research on Earthquake Engineering (Taiwan) was also acknowledged.

## REFERENCES

1. Wen, Y.K., “Method of random vibration of hysteretic systems,”, J. Eng. Mech. ASCE 102 (2), 1976, p. 249–263.
2. Ikhouane, F.. V. Manosa, and J. Rodellar, "Dynamic properties of the hysteretic Bouc-Wen model," Systems and Control Letters, 56:3, 2007, p:197-205.
3. Doebling, S.W., Farrrar, C.R. and Prime, M.B, “A summary Review of Vibration-based Damage Identification Methods,” shock Vibration digest, 30(2), 1998, p:91-105.
4. Yen, G.G. and K.-C. Kuo, “Wavelet packet feature extraction for vibration monitoring,” IEEE Transactions on Industrial Electronics, 47 (3): 2000, p. 650-667.
5. Sun, Z. and Chang, C. C., ”Structural damage assessment based on wavelet packet transform”, ASCE, J. Struct. Eng., 128(10),2002, p1354-1361.
6. Han, J.G., W.X. Ren, and Z.S. Sun, “Wavelet packet based damage identification of beam structures,”, International Journal of Solids and Structures, 2005. 42 (26): p. 6610-6627.
7. Uang, C. M. and Bertero, V. V., “Evaluation of Seismic Energy Structures,” Earthquake Engineering and Structural Dynamics, Vol.19, 77-90, 1990.
8. Alonso, F.J., J.M. Del Castillo, and P. Pintado, “Application of singular spectrum analysis to the smoothing of raw kinematic signals,” Journal of Biomechanics, 2005. **38**(5): p. 1085-1092.
9. Park, Y.-J. and A. H. S. Ang, "Mechanistic seismic damage model for reinforced concrete," Journal of Structural Engineering, 111:4 (1985), 722-739.
10. Park, Y.-J., A. H. S. Ang, and Y. K. Wen, "Seismic damage analysis of reinforced concrete buildings," Journal of Structural Engineering, 111:4 (1985), 740-757.
11. Robertson, A.N., C.R. Farrar, and H. Sohn, *Singularity detection for structural health monitoring using Holder exponents.* Mechanical Systems and Signal Processing, 2003. **17**(6): p. 1163-1184.

Proceedings of the IMAC-XXVIII
February 1–4, 2010, Jacksonville, Florida USA


# Composite Damage Detection Using Laser Vibrometry with Nonlinear Response Characteristics

Sara S. Underwood and Douglas E. Adams
Purdue University, School of Mechanical Engineering, Center for Systems Integrity
1500 Kepner Drive, Lafayette, IN, USA 47905

## ABSTRACT

Composite materials are susceptible to subsurface damage mechanisms (delaminations, disbonds, etc), which are difficult to detect visually. Current inspection techniques for composite materials used in military aircraft are time consuming and typically require removal of the component for inspection. In this paper, a technique is discussed that uses multi-amplitude frequency response measurements collected with a three-dimensional scanning laser vibrometer to detect and locate damage by observing nonlinear vibration response characteristics introduced into composite specimens by material damage. A carbon fiber composite panel with several impact damage locations is investigated. The panel is excited by a piezoelectric actuator and the surface velocity is measured with a 3-D scanning laser vibrometer. Nonlinear characteristics in the measured response at multiple amplitudes of excitation are used to locate and detect the subsurface damage introduced by the impacts. The higher energy level impacts investigated were successfully detected using this reference-free technique. In addition, the damage on the panel is further localized by investigating nonlinearities at higher frequency ranges.

## INTRODUCTION

Composite materials are susceptible to subsurface damage mechanisms, which are difficult to detect visually. These damage mechanisms include delaminations, disbonds, and core cracking or crushing. Current inspection techniques used in military aircraft are time consuming and typically require removal of the component for inspection. A robust technique is needed to inspect composite materials for subsurface damage that addresses these issues. In this paper, a technique is discussed that uses a three-dimensional scanning laser vibrometer to measure the forced frequency response of a carbon fiber composite panel as it is excited by a piezoelectric actuator. Analysis of the frequency response functions obtained at varying excitation amplitude levels is performed to identify nonlinear vibration response properties introduced to the panel by material damage. The long-range capability of the scanning laser vibrometer allows for wide-area inspection of a composite specimen, therefore eliminating the need for an aircraft component to be removed for inspection.

In this paper, a reference-free damage detection approach previously developed in [1] was applied to a carbon fiber composite panel to investigate the damage detection capability for more realistic damage mechanisms. In [1], damage mechanisms such as core cracking and disbonds were investigated on fiberglass composite specimens where the damage was introduced by placing cuts in the core or between the core and face sheets of the specimen investigated. In this paper, impact damage was applied to a carbon fiber composite panel in order to introduce damage which was more representative of what is seen in reality. The scanning laser vibrometer is used to measure the surface velocity of the panel to a sine sweep excitation and the resulting frequency response functions produced are analyzed to determine locations on the panel displaying the largest nonlinear response properties. Several different levels of damage are investigated at various locations on the panel, and thermal images of a similar panel are shown to confirm the damage indices produced through the measurement and analysis obtained with the scanning laser vibrometer. In addition, it is shown that investigation of nonlinearities at increasingly higher frequency ranges allows for better localization of the damage.

**EXPERIMENTAL INVESTIGATION**

Impact damage on a carbon fiber sandwich panel was investigated experimentally using a three-dimensional scanning laser vibrometer. The forced frequency response of the panel to a sine sweep excitation was measured at specified points on the panel, and nonlinear response characteristics were identified. The areas on the panel displaying the largest nonlinear response characteristics were indicative of subsurface damage.

The panel investigated consisted of carbon fiber face sheets sandwiching an Aramid honeycomb core. A diagram depicting the layout of the panel is shown in Figure 1. This diagram shows the four carbon fiber plies on either side of the honeycomb core, as well as the orientations of the fibers in each ply. A description of the material used for each layer along with the ply angle and thickness is given in Table 1. The panel was cut to a length of 26 in and a width of 12 in.

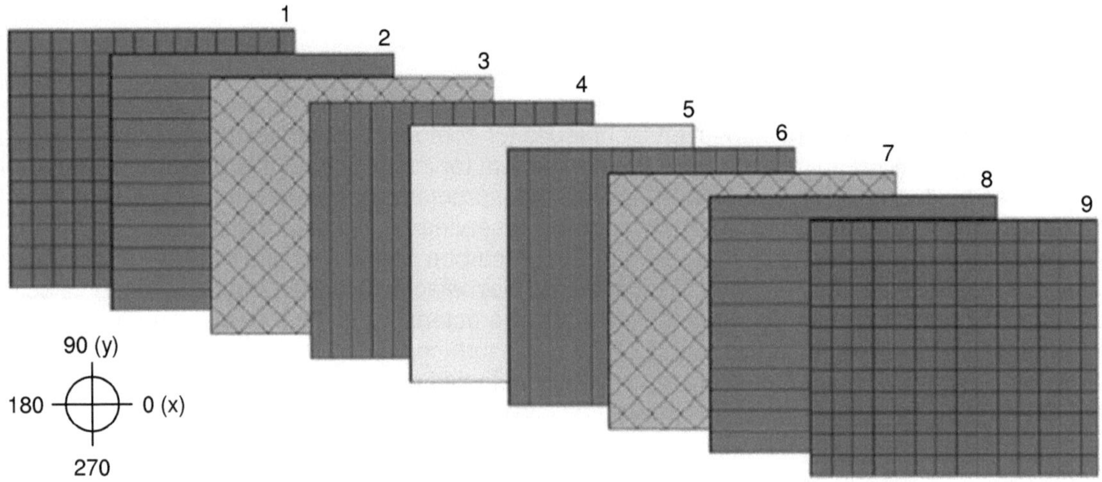


Figure 1. Diagram of the panel lay-up for the carbon fiber composite panel.

Table 1. Description of each layer of the carbon fiber composite panel.

| Layer Number | Description | Ply Angle (degrees) | Thickness (in) |
|---|---|---|---|
| 1 | 200 gsm 3K Carbon 2x2 Twill | 0 | 0.00866 |
| 2 | 300gsm Std. Modulus Carbon Uni T700 | 0 | 0.01180 |
| 3 | 600gsm Carbon/Epoxy Sprint | 45 | 0.02360 |
| 4 | 300gsm Std. Modulus Carbon Uni T700 | 90 | 0.11800 |
| 5 | Aramid Honeycomb 1/8-3lb | 0 | 0.50000 |
| 6 | 300gsm Std. Modulus Carbon Uni T700 | 90 | 0.11800 |
| 7 | 600gsm Carbon/Epoxy Sprint | 45 | 0.02360 |
| 8 | 300gsm Std. Modulus Carbon Uni T700 | 0 | 0.01180 |
| 9 | 200 gsm 3K Carbon 2x2 Twill | 0 | 0.00866 |

In order to introduce damage to the panel at controlled levels, an Instron Dynatup impact tower (model 9259HV) was used to impact the panel at five locations. Varying impact energies of 2, 3, 5, and 7 ft-lb were used. Figure 2(a) shows the locations and energies of the impacts on the panel. The impact damage introduced face sheet cracking and core crushing at the point of impact, as well as delamination around the point of impact. The severity of the damage at the point of impact and the size of the delamination varied mainly based on the impact energy. Whereas in [1] where the damage mechanisms investigated were simulated disbonds or core cracking, here, the damage introduced by the impacts is more realistic and the damage mechanisms are coupled together.

A PCB piezoelectric actuator (model 712A02) with a PCB impedance head (model 288D01) was attached at a skewed angle to the bottom left hand corner of the panel in order to excite the panel and measure the input force to the panel. The actuator location is highlighted in Figure 2(a). A sine sweep excitation from 100 to 5000 Hz was used for the excitation signal with a sample time of 1.28 s. Various excitation amplitudes were used to excite the panel. A grid of measurement points was created about each damage location through the scanning laser vibrometer software. These grids are shown in Figure 2(b). Each grid consisted of 25 points and was centered about the point of impact. In addition, each grid covered an area of approximately 16 $in^2$, giving about 1 inch spacing between grid points.

The scanning laser vibrometer measures the surface velocity of the panel at each grid point in three orthogonal directions. The frequency response function relating the input force to the panel with the output surface velocity in the transverse, lateral, and longitudinal directions is computed within the scanning laser vibrometer software. One sample was taken for each sine sweep excitation, and twenty-five averages were collected for each measurement point. Consecutive scans at different excitation amplitude levels were performed for each damage location.

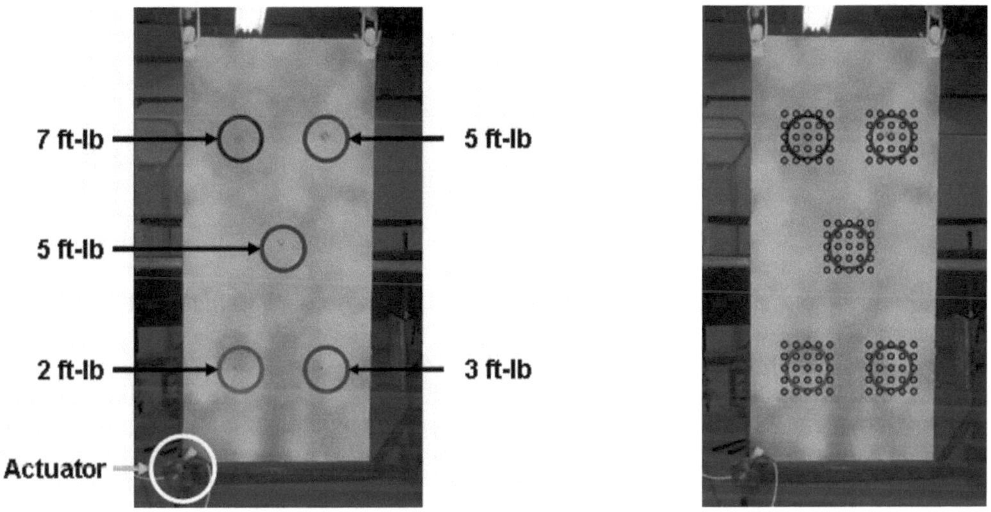


(a) (b)

Figure 2. Schematics showing (a) the impact locations and energies and (b) the corresponding grids used for measurement with the scanning laser vibrometer.

## ANALYSIS THROUGH NONLINEAR IDENTIFICATION

The frequency response functions obtained from varying amplitude excitation levels were compared in order to identify locations displaying nonlinear response properties to the sine sweep excitation. As described in [1] for fiberglass composite panels, the areas displaying the largest nonlinear response properties are indicative of subsurface damage. This same analysis approach was applied to the carbon fiber panel to investigate the behavior of the damage introduced by the impacts. The magnitudes of the frequency response functions obtained from a high and a low amplitude excitation were compared by computing the absolute value of the difference between the two frequency response functions. A summation of this difference across the frequency range from 315 to 850 Hz was computed for each measurement direction and a damage index was then created based on the product of the results in the transverse, lateral, and longitudinal directions. The measured points displaying larger nonlinear response properties were assigned higher damage index values.

Unlike the fiberglass panels investigated in [1], which were very highly damped, the carbon fiber panel investigated in this paper was lightly damped. Because of the low damping, the higher excitation frequencies were not damped out, allowing a response to be obtained across the entire frequency range excited. In addition, care had to be taken to ensure the panel was not overexcited, particularly at locations close to the excitation source. Excitation amplitudes that were too high at locations close to the excitation source introduced nonlinearities that were not due to material damage, bringing error into the damage index results. To determine the optimal excitation levels for this analysis, an amplitude study was performed at the 5 ft-lb impact location at the top of the panel. The damage indices produced for different combinations of input voltages to the actuator are shown in Figure 3. The approximate location of the impact is highlighted in each damage index by the small black circle. Due to the coarse nature of the 25 point grids used in the measurement, interpolation was used between grid points to obtain a continuous damage index.

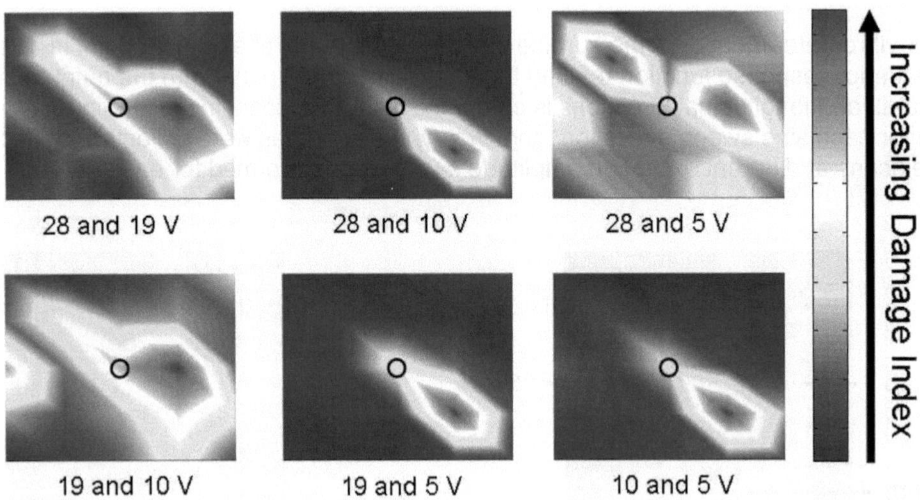


Figure 3. Results of amplitude study performed at the upper 5 ft-lb impact location.

As seen from Figure 3, the choice of high and low amplitude excitation levels had a large effect on the damage index produced. However, it is noted that the areas in the damage indices indicating the highest damage levels are located in the vicinity of the impact damage (taking into account the coarse nature of the grids used), which is where delamination between the carbon fiber plies occurred. The results of the amplitude study suggested that the largest spread between high and low amplitude excitation levels was optimal. However, further tests performed at the other impact locations closer to the excitation source limited the high amplitude excitation level, due to excessive excitation of the panel. The maximum excitation amplitude that was could be used at the 2 ft-lb and 3 ft-lb impact locations was a 10 V input to the actuator.

The ordinary coherence for the high and low amplitude frequency response functions was checked to ensure that the surface velocity output was correlated with the input force for the measurements being performed. The coherence in the transverse direction was near unity across the excitation range except where antiresonances occurred in the frequency response functions, as expected. In the lateral and longitudinal directions, the coherence was less than unity across the excitation range due to the small responses measured in those in-plane directions. However, the coherence increased at peaks in the frequency response functions suggesting good correlation at frequencies where a measurable response was produced.

## RESULTS

Damage indices were produced for each of the impact locations using the measurement grids shown in Figure 2(b). The results were overlaid on the damage locations and are shown in Figure 4. The high and low amplitude excitation levels used were 10 V and 5 V, respectively. These amplitudes were chosen in order to be consistent with the results for all of the damage locations. It is noted that the damage indices were all created individually for each impact location, and do not reflect the severity of damage relative to each other.

As seen in Figure 4, the damage index results in each case indicate that the point located directly on the impact location (the point at the center of each damage index) does not display large nonlinear response properties when a change in excitation amplitude is experienced. This is most likely due to the cracking and change of shape of the top face sheet at the impact location. For the three impact locations on the upper half of the panel, including the two 5 ft-lb impacts and the 7 ft-lb impact, the delamination surrounding the point of impact is apparent in the computed damage indices. The interpolation used in the creation of the damage indices resulted in a less than desirable image of the damage shape and size. However, the result showed that sub-surface damage is present in the impacted region of the panel. If more information about the overall shape and size of the damage was desired, a finer grid could be created to obtain those results.

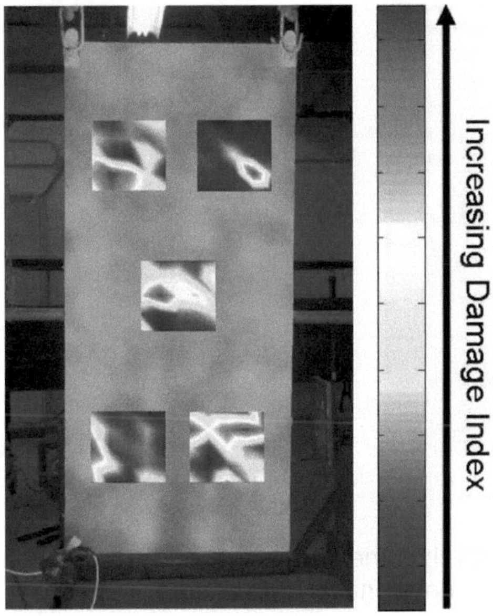


Figure 4. Damage indices for each impact overlaid on the respective damage locations.

The damage indices created for the 2 and 3 ft-lb impacts located on the bottom half of the panel do not clearly indicate damage from the impacts. This is most likely a result of the low damage levels compared to the 5 and 7 ft-lb impacts, the close proximity of the damage to the actuator location, or a combination of both of these reasons. As described previously, the close proximity of these damage locations to the excitation source may have caused nonlinearities introduced by material damage to be washed out by nonlinearities introduced through the response of the panel to a large excitation. Measurement points experiencing large deflections may falsely indicate damage through the analysis method performed. In order to more accurately identify the damage introduced by the impact damage to this section of the panel, it would be necessary to reduce the amplitude of the excitation so that fewer nonlinearities are introduced by the actuator overexciting the panel. Another option would be to move the actuator to another location on the panel, such as the upper right corner of the panel. This second actuator location, if coupled with the results from the current actuator location, would allow for more accurate identification of the damage introduced by all of the impact locations investigated.

In order to justify the results obtained and shown in Figure 4, thermal images were produced for a second carbon fiber panel, cut to the same dimensions as the panel used in the scanning laser vibrometer investigations, with the same impact locations and energies as is depicted in Figure 2(a). An infrared camera was used to obtain the thermal images, providing visual images of the shape and scope of the subsurface damage in the panel. The images obtained for each of the damage locations are shown in Figure 5(a). In addition, Figure 5(b) shows a clearer image of the delamination around the 7 ft-lb impact on this second panel.

It is clear from Figure 5(a), that both 5 ft-lb impacts and the 7 ft-lb impact produced visible subsurface damage in the region surrounding the impact location. However, very little, if any, subsurface damage is seen at either the 2 ft-lb or the 3 ft-lb impact locations. This is most likely a reason why the scanning laser vibrometer analysis produced poor results at these two damage locations. A seam in the panel running down the right side of the panel can also be seen in the images produced for both 5 ft-lb impacts as well as the 3 ft-lb impact. This is a manufacturing defect which could also affect the results obtained through measurement with the scanning laser vibrometer. Figure 5(b) clearly shows the delamination resulting from the 7 ft-lb impact. The blue region in the immediate area of the impact suggests subsurface damage near the surface, while a yellow region can be seen extending above and below this blue region. This shows a deeper delamination extending even further from the point of impact. The peanut shaped behavior of these delaminations comes about from the stacking sequence of the carbon fiber plies.

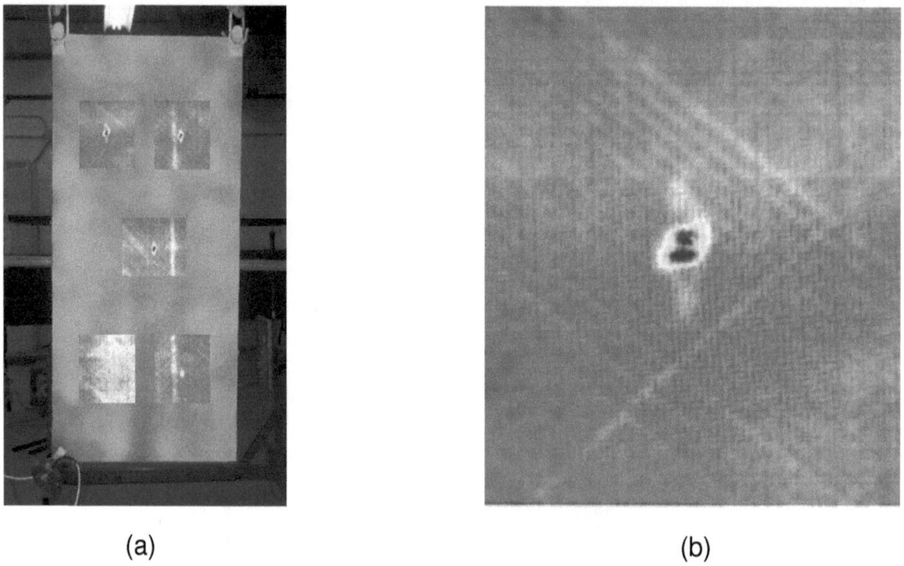


Figure 5. Thermal images of (a) each of the impact damage locations and (b) the 7 ft-lb impact showing the delamination produced by the impact.

Similar research on composite specimens has shown that higher modal frequencies may be used to more accurately detect and locate damage [2,3,4]. This was investigated on the first carbon fiber panel with the scanning laser vibrometer for the 5 ft-lb impact in the upper corner of the panel. The original range used to create a damage index for this impact location was from 315-850 Hz. In order to see the effects of higher frequency ranges, additional damage indices were created using the summation of the difference between high and low amplitude excitation levels for frequency ranges of 850-1500 Hz and 1500-2500 Hz. The damage indices for these three frequency ranges are shown in Figure 6. This analysis is consistent with the findings in [2-4] in that the damage becomes more localized at higher frequencies. In addition, the error in the measurement seems to decrease as the frequency range is increased. As seen in Figure 6, the frequency range from 1500-2500 Hz gives the clearest indication of damage about the point of impact, whereas the frequency range from 315-850 Hz provides a less clear image of the damage location.

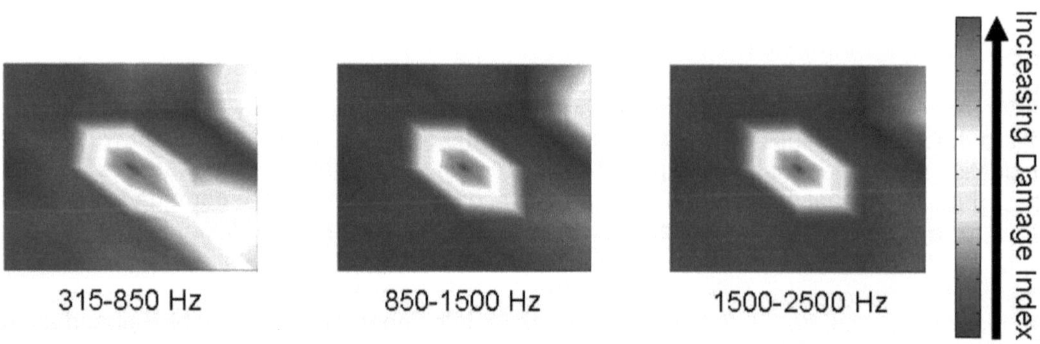


Figure 6. Damage indices produced at increasingly higher frequency ranges at the upper 5 ft-lb impact location.

## DISCUSSION

The analysis and results obtained for the detection of subsurface damage in the carbon fiber composite panel through investigation of nonlinear frequency response properties provided insight into the ability to detect and locate subsurface damage through this technique. Subsurface damage introduced by impacts of 5 and 7 ft-lb was accurately detected by analysis of frequency response functions measured through a scanning laser vibrometer at several excitation amplitudes. It was found that the 2 ft-lb and 3 ft-lb impacts located close to the actuator were not able to be accurately detected at the excitation amplitudes used for the investigation. This could be a result of the low level of damage introduced by the impacts as was seen through thermal images of a similar panel, or it could be a result of overexciting the panel at locations close to the excitation source. Reducing the amplitude of the excitations used may improve the results at these locations close to the actuator, as well as using a second actuator location at a different location on the panel. The overall results of this investigation showed that the ability to detect and locate subsurface damage is sensitive to many factors, including the amplitudes used for the high and low amplitude excitation levels, the actuator location in proximity to the damage location, the severity of the damage, and the frequency range of used for the excitation and for performing the analysis. Frequency ranges higher than those investigated here may be achieved on the carbon fiber panel due to its low damping properties, which may allow for even better detection and localization of the damage.

## CONCLUSION

A reference-free damage detection approach was investigated on a carbon fiber composite panel using a three-dimensional scanning laser vibrometer. Impact damage was introduced to the panel at varying impact energies and at various locations. The scanning laser vibrometer was used to measure the surface velocity of the panel to a sine sweep excitation. Frequency response functions were produced relating the surface velocity in the transverse, lateral, and longitudinal directions to the input force to the panel from the excitation. Analysis of the frequency response functions at multiple amplitudes of excitation was performed to identify regions of the panel displaying nonlinear frequency response properties. This allowed for damage indices to be produced. Subsurface damage introduced by two 5 ft-lb impacts and a 7 ft-lb impact was successfully detected and confirmed by thermal imaging of a similar carbon fiber panel. 2 ft-lb and 3 ft-lb impacts were also investigated but were unable to be accurately detected through the nonlinear analysis technique. Several reasons for the inability to detect these low impact energy levels were discussed. In the case of one of the 5 ft-lb impacts, investigation of nonlinearities at increasingly higher frequency ranges allowed for better localization of the damage.

## REFERENCES


[1] Underwood, S., D. Adams, D. Koester, M. Plumlee, and B. Zwink. "Structural damage detection in a sandwich honeycomb composite rotor blade material using three-dimensional laser velocity measurements," presented at the American Helicopter Society 65th Annual Forum, May 27-29, 2009.

[2] Vanlanduit, S., P. Guillaume, J. Schoukens, and E. Parloo. "Linear and nonlinear damage detection using a scanning laser vibrometer." Proceedings of SPIE, Vol. 4092, 453-466, 2000.

[3] Ghoshal, A., A. Chattopadhyay, M.J. Schulz, R. Thornburgh, and K. Waldron. "Experimental investigation of damage detection in composite material structures using a scanning laser vibrometer and piezoelectric actuators." Journal of Intelligent Material Systems and Structures, Vol. 14, No. 8, 521-537, 2003.

[4] Budde, C., S. Underwood, D. Koester, and D.E. Adams. "Impact load estimation and damage detection for fiberglass composite rotor blades," presented at the 7th International Workshop on Structural Health Monitoring, September 9-11, 2009.

MIX
Papier aus verantwortungsvollen Quellen
Paper from responsible sources
FSC® C105338

If you have any concerns about our products,
you can contact us on
**ProductSafety@springernature.com**

In case Publisher is established outside the EU,
the EU authorized representative is:
**Springer Nature Customer Service Center GmbH**
**Europaplatz 3, 69115 Heidelberg, Germany**

Printed by Libri Plureos GmbH
in Hamburg, Germany